£ 4-60

WEATHERING THE
WILDERNESS

The Sierra Club Outdoor Activities Guides

WEATHERING THE WILDERNESS

The Sierra Club Guide to Practical Meteorology

William E. Reifsnyder

Sierra Club Books   San Francisco

The Sierra Club, founded in 1892 by John Muir, has devoted itself to the study and protection of the earth's scenic and ecological resources—mountains, wetlands, woodlands, wild shores and rivers, deserts and plains. The publishing program of the Sierra Club offers books to the public as a nonprofit educational service in the hope that they may enlarge the public's understanding of the Club's basic concerns. The point of view expressed in each book, however, does not necessarily represent that of the Club. The Sierra Club has some sixty chapters coast to coast, in Canada, Hawaii, and Alaska. For information about how you may participate in its programs to preserve wilderness and the quality of life, please address inquiries to Sierra Club, 730 Polk Street, San Francisco, CA 94109.

Library of Congress Cataloging in Publication Data

Reifsnyder, William E.
Weathering the wilderness.

Includes index.
1. Weather. 2. Hiking. I. Sierra Club. II. Title.
QC981.45.R44   551.5′02′47965   79-20859
ISBN 0-87156-266-9

Jacket illustration by Jon Goodchild
Book design by Mark Jacobsen
Illustrations by Craig DuMonte

Printed in the United States of America on acid-free recycled paper
10 9 8 7 6 5 4

We gratefully acknowledge the following for permission to reprint copyrighted material: Sidney Licht, M.D., Elizabeth Licht, Brooks E. Martner, Richard A. Dirks, Val Eichenlaub, Thomas Hodler, Robert Muller.

Contents

Acknowledgments

Many people helped me in the preparation of this book. Without the unselfish dedication of hundreds of cooperative observers of the national weather services of the United States and Canada, climatic information for most of our remote wild areas would simply not be available. But I have also had the cooperation of numerous officials of the U.S. Climatic Data Service and the Canadian Climate Center, and their help is also gratefully acknowledged.

In particular, I wish to thank several of my associates who read early drafts and, by their critical reviews, contributed to the organization of the regional climatologies: Douglas Burbank, Robert Daniels, Edward Garvey and Joel White.

Several State Climatologists were especially helpful in providing hard-to-find climatic summaries. These include Robert Lautzenheiser (New England); A. Boyd Pack (New York); Robert O. Weedfall (West Virginia); Orman E. Street (Maryland); Earl L. Kuehnast (Minnesota); and Fred V. Nurnberger of the Michigan Weather Service. G. A. McKay, Director, Climatological Applications Branch, Canadian Climate Center, provided help with Canadian data.

Brooks E. Martner and Richard A. Dirks of the University of Wyoming provided unpublished data on the climate of Yellowstone National Park, as did Mary M. Reynolds for Olympic National Park. For permission to use material originally contained in my chapter, "What is Weather" in *Medical Climatology* (New Haven: Elizabeth Licht Publisher), I am indebted to the late Dr. Sidney Licht, editor of the volume.

Finally (although this really should be first), the book would not have come into being without the encouragement, critical reading, and editorial assistance of Marylou Reifsnyder. To her I dedicate this book.

PART I

An Introduction to Weather

Chapter 1

The Air
Around Us

We live at the bottom of the ocean of air that is the earth's atmosphere. Although most of the time we are scarcely aware of its existence, we are completely dependent on the atmosphere for the chemicals that sustain life: oxygen, carbon dioxide, and water. Indeed, we are so dependent on the sea-level atmosphere that we cannot exist indefinitely at altitudes exceeding about 18,000 feet, where the lung contains only about half the number of molecules it contains at sea level. We are made aware of this as our breathing becomes more labored when we hike or climb in the high mountains.

The atmosphere is a mixture of gases; it contains varying amounts of suspended liquid and solid particles. Some of these components do not change much from day to day or from place to place over the earth's surface. Oxygen, for example, constitutes about 20 percent of a volume of dry air, and this proportion does not change significantly with elevation, location, or time, except where rapid oxidation is occurring. Nitrogen is even more invariant because of its relative inertness; it constitutes nearly all the remainder of the gaseous content of dry air. A considerable number of other gases, mostly inert, are present in small quantities.

Carbon dioxide and water, the raw materials of photosynthesis, are present in varying amounts. Water vapor can constitute up to about 5 percent of the mass of a volume of air. The normal concentration of carbon dioxide is only .03 percent, but this is subject to considerable variation near sources of carbon dioxide (in regions of active combustion, respiration, or oxidation) and near regions of carbon dioxide utilization in photosynthesis.

Ozone, the triatomic form of oxygen, is present in the atmosphere in very small amounts. Its importance to the maintenance of tolerable conditions for life, however, is very great. Like carbon dioxide and water vapor and unlike nitrogen and oxygen, ozone is opaque to certain specific wave lengths of radiant energy. Ozone absorbs, at high altitudes, the bulk of the ultraviolet radiation from the sun, preventing it from reaching the ground. Water vapor and carbon dioxide are relatively transparent to the shortwave radiation from the sun but absorb much radiation in the infrared region.

Various other gases are also present in the atmosphere, generally in insignificant amounts but occasionally in amounts that are biologically important. Products of incomplete combustion, effluents of various industrial processes, and evaporation of plant-produced volatile hydrocarbons contribute to the total content of any volume of atmospheric air.

The atmosphere is also the vehicle for quantities of solid and liquid particles. Water, the most variable of all atmospheric constituents, can exist in liquid and solid state as well as in its gaseous form. Dust blown from the surface, volcanic ash, carbon particles from incomplete combustion, meteoritic dust, residue from nuclear explosions, and salt particles from the sea all contribute to the composition of natural air. These tiny particles settle so slowly that they remain in solution in the atmosphere as aerosols. Not all are undesirable impurities: certain types of particles are essential to condensation and precipitation processes. Distribution of these aerosols varies widely over the surface of the earth. Average concentrations (excluding water droplets and ice particles) may range from about 1,000 per cubic centimeter over open ocean to more than 150,000 per cubic centimeter over cities.

Atmospheric Pressure

The molecules that make up the atmosphere have mass, which is under the influence of the earth's gravitational field.

Figure 1-1. Variation of pressure, temperature and density with height

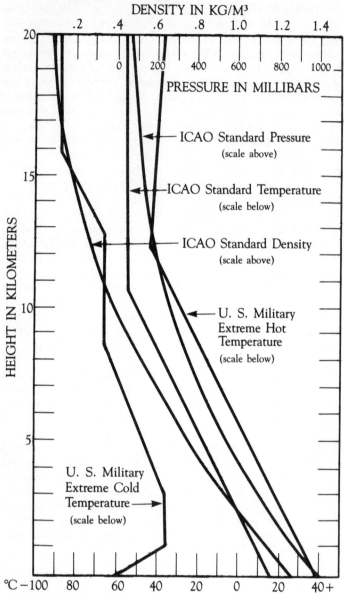

Standard conditions as specified by the International Civil Aviation Organization (ICAO) and the U. S. Army.

Therefore, they have weight. Immersed in this sea of air as we are, we cannot sense this weight directly or even measure it with an ordinary balance. But weigh it does, and it presses on the earth's surface at sea level with a force of about 15 pounds per square inch. Masses of air near sea level are compressed more than those at higher levels because the entire depth of atmosphere is pressing down on them. Thus, there are more molecules in a unit volume of air at sea level than there are at the top of a high mountain. At 18,000 feet, the density of air is only about half that at sea level. The relative proportion of the invariant components does not change, but the absolute number of molecules per unit volume does. The average variation of pressure and density with height is indicated in figure 1-1.

The total weight, or pressure, of the atmosphere at any level is the sum of the weights or pressures exerted by the quantity of each constituent gas from that level to the top of the atmosphere. Each gas exerts a partial pressure, the sum of which equals the total pressure at that level. In order to calculate the partial pressure of a particular gas, it is necessary to know only the total pressure and the mass proportion of that gas in the total air column above the point. For oxygen, the proportion can be considered constant at about 20 percent. For water vapor, the proportion varies with height; the proportion is generally highest near the surface, the ultimate source of all atmospheric moisture.

Although the vertical variation of pressure is much greater than the horizontal variation, it is the horizontal variation that gives rise to the winds and great wind systems of the globe. In regions heated by the sun, low pressures develop, relative to cooler regions nearby. Horizontal pressure gradients develop, and air tends to move from regions of higher pressure to those of lower pressure. The maximum variation of pressure at sea level is only about 5 percent from the mean value of 1,013 millibars (29.92 inches of mercury); and horizontal gradients over large distances rarely exceed 50 millibars per hundred miles, even in tropical hurricanes. By contrast, a 5 percent reduction from sea-level pressure occurs at an elevation of about 2,000 feet.

At a particular location, atmospheric pressure varies primarily as a result of the migration of large systems of high and low pressure. A smaller diurnal fluctuation of pressure occurs as the result of atmospheric tides. Although this fluctuation is frequently lost in the larger variations caused by moving systems, it is nevertheless present and can be detected. In areas where the larger variations are not common, the diurnal fluctuation is readily observed; it may amount to about 3 millibars in equatorial regions.

Air Temperature

The temperature of a body or of a gas is a manifestation of the energy level and motion of the constituent molecules. Temperature and heat are not synonymous; two apparently similar bodies at the same temperature may have different heat contents. The differentiating parameter is the heat capacity, that quantity of heat required to raise the unit mass of a substance a unit temperature difference. The physical importance of temperature is that it controls the direction of heat flow: heat invariably flows from a region of high temperature to a region of low temperature in the absence of work performed on the system.

Temperature scales are referred to fixed levels. The melting point of pure ice is one of these; absolute zero, the point where all molecular motion ceases, is another. The boiling temperature of water depends on the pressure of the atmosphere, decreasing as the pressure decreases, and is therefore not a good reference point unless the pressure is measured and the boiling point for that pressure is known. Indeed, this dependence of the boiling temperature and air pressure is the basis for a classic method of

Table 1-1.   Boiling point of water and cooking time for food at various elevations

Altitude (feet)	Boiling temperature (°F)	(°C)	Cooking time (minutes)
Sea level	212°	100°	10
1,000	210°	99°	11
2,000	208°	98°	12
5,000	203°	95°	15
7,500	198°	92°	18
10,000	194°	90°	20
15,000	185°	85°	25

For a food that takes longer to cook at sea level, form a simple proportion. Thus, a twenty-minute cooking time at sea level doubles all the other times: twenty-two minutes at 1,000 feet, and so on.

measuring altitude. Early explorers measured the temperature of boiling water to estimate their elevation above sea level. For the hiker, the main significance is that cooking times of boiled foods increase dramatically with elevation. Table 1-1 indicates the boiling temperature of water at various elevations and the approximate cooking times for a food that takes ten minutes to cook at sea level.

Humidity

As indicated previously, water is a variable constituent of air. If pure water is maintained at a fixed temperature in a closed system from which air is evacuated, some of the water will evaporate, or boil, filling the empty space. An equilibrium will eventually be reached between the water vapor and the liquid water. The pressure exerted by the gaseous water vapor will be constant for a particular temperature of the water, increasing as the water temperature increases. If the same experiment is performed with the air at normal atmospheric pressure above the water to begin with, equilibrium will be reached when the same amount of water vapor is present in the air, although the approach to equilibrium will be much slower. The pressure exerted by the water vapor portion of the air and water vapor mixture will be the same as that exerted by the water vapor alone in the evacuated system. This equilibrium pressure, the partial pressure of the water vapor, depends solely on the temperature of the water with which it is in equilibrium. This is defined as the saturation vapor pressure of the water vapor at the particular temperature.

However, air is usually not saturated; the partial vapor pressure of the contained water is lower than the saturation vapor pressure at the temperature of the air. The ratio of actual vapor pressure of an air parcel relative to the saturation vapor pressure is known as the relative humidity; it is usually expressed as a percentage. Because the saturation vapor pressure increases as the temperature increases, the relative humidity of a parcel of air that is being heated without the addition or subtraction of moisture will decrease. Thus, the common occurrence of a diurnal cycle of relative humidity is the opposite of the temperature cycle: as the temperature goes up, the humidity goes down, even though the water vapor content of the air does not change.

A typical trace of the temperature and relative humidity during a twenty-four-hour period of clear weather during which there was no change in the actual amount of water vapor in the air is shown in figure 1-2.

Figure 1-2. Typical diurnal variation of air temperature and relative humidity

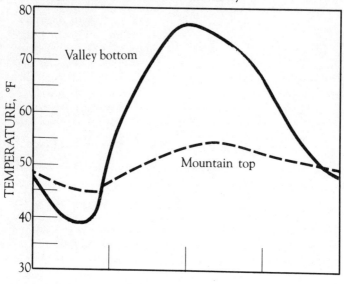

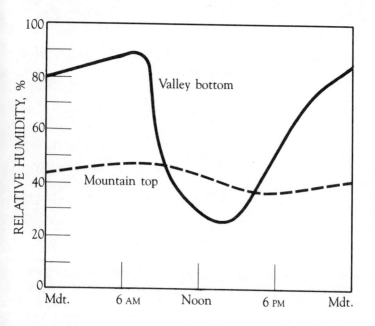

For the recreationist, the relative humidity of the atmosphere is in many ways more important than the absolute humidity or the actual amount of water vapor in the atmosphere. The reason for this is that the moisture content of many natural materials depends on the relative humidity of the surrounding air. Dead grass is one such material: at low relative humidities, below about 20 percent, dead grass is very dry and highly flammable. As the humidity increases, the grass takes up moisture (even though the actual amount of water vapor in the air does not change) and becomes much less flammable. In the early morning, with the humidity near 100 percent, it may be difficult to get grass and small twigs to burn; in the afternoon, a burning cigarette may be enough to ignite fine dry grass.

This relationship between relative humidity and the moisture content of biological materials is used in the construction of the common hair hygrometer. As the humidity increases, the hair absorbs moisture from the air and stretches. By connecting one end of a bundle of human hair to a pointer moving over a calibrated dial, one can construct an instrument for measuring relative humidity.

Atmospheric humidity can also be measured with a mercury thermometer with a moistened wick surrounding the mercury reservoir. With adequate ventilation of the wick, its temperature is lowered by evaporation to a constant value, the wet-bulb temperature. This, together with the air temperature and suitable tables, can be used to determine the dew point or the relative humidity. The wet-bulb temperature has biological significance in that it is the lowest temperature to which a body can be cooled by evaporating water from it into a moving airstream. Figure 1-3 shows the average wet-bulb temperature over the United States in July.

Wind Circulation

If we were to release into the atmosphere a balloon so balanced it would neither rise nor fall but would faithfully follow the air currents and if we could follow this balloon over the course of a day, we would be taken on a merry chase indeed. Much of the time it would be carried on a path dictated by the horizontal pressure gradients associated with large-scale weather systems, rising or falling only slowly but scudding along with the speed of the wind. At other times, it might be caught in a rapidly rising vertical current of air, in a giant thunderhead, only to be brought violently back down in a cold downdraft. During the night, it

Figure 1-3. Average July wet-bulb temperatures in the United States

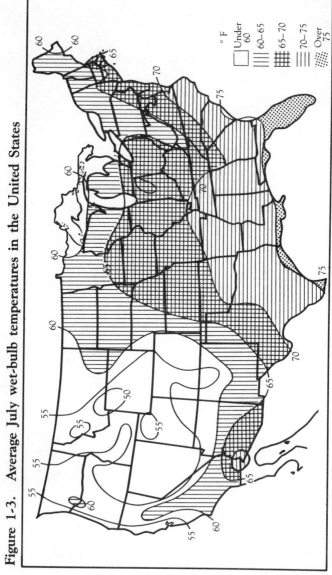

Based on 200 first-order weather bureau stations, 1899–1938 (Source: U. S. Dept. of Commerce, National Weather Service)

might be found drifting slowly down a canyon, tracing the path of the cold air draining off the canyon walls and winding its way down the twists and turns of the valley.

The air movement made visible by the meanderings of the balloon might seem at first glance completely random and chaotic. But, of course, the motions are not random; they result from highly patterned and structured pressure gradients in the atmosphere. To understand these motions, we must first see how and why the pressure patterns develop.

On a sunny summer day, air heated by contact with the hot ground tends to rise. It does so in organized "thermals," columns of upward-moving air that are sought by hang-glider pilots and soaring birds alike. If there is sufficient moisture in the air, the cooling caused by the expansion of the rising air will eventually lead to condensation and the formation of a cumulus cloud. The puffy, fair weather cumulus cloud represents the top of such a rising column of air; the flat base is the level at which condensation is occurring. The extra energy provided by the condensing water vapor may push the air to great heights, triggering a thunderstorm.

If the heated air lies over a large, flat area, such as a desert playa, the upward convection, perhaps triggered by the random jumping of a desert jackrabbit, may organize itself into the tight whirl of a dust devil. Such dust devils may wander for hours over the desert, feeding on the vast reservoir of heated air near the hot surface, and may die out only when the sun sinks low in the sky, denying the ground surface of its source of heat energy.

But no matter what the cause of the thermal imbalance, these vertical motions rob the surface of some of its air, thus creating a region of relatively low pressure. A horizontal pressure gradient develops and air moves horizontally to redress the balance.

For every upward current, there must be a compensating downward movement of air. These downdrafts are normally spread over larger areas and thus have downward velocities that are much smaller than the updrafts. Even in a region experiencing widespread active thunderstorms, the area of downward-moving air between storms may be many thousand times as large as the area of updrafts in the storms. This sinking motion of a large mass of air is known as subsidence. Subsidence is a feature of surface high-pressure areas and is usually associated with clear skies and stable conditions.

Horizontal wind.   In order to complete the cycle of air movement in the atmosphere, horizontal winds are necessary to join the upward- and downward-moving currents. It is the uneven heating of the earth's surface that produces our winds by first producing vertical currents. As seen on a weather map, horizontal winds appear to be caused by pressure gradients and are not obviously related to variations in surface heating. But ultimately, the motions may be traced back to this source: the energy received from the sun and absorbed in an uneven pattern over the variegated surface.

However, it is upon the large-scale pressure patterns that we must focus our attention in order to understand the nature of our global wind system. If we study a surface weather map showing the distribution of winds and pressures, we will notice that in regions other than the tropics, the winds do not blow directly from regions of high pressure to centers of low pressure. In the Northern Hemisphere they spiral counterclockwise into the lows. At high levels, air moves scarcely at all in the direction the gradient of pressure would suggest. It moves perpendicular to the gradient, that is, around regions of low and high pressure, but not to or from them.

This seeming paradox is explained by the earth's rotation around the polar axis. Because of this rotation, any object given a push in some direction experiences an apparent sideways push, to the right of its motion in the Northern Hemisphere. This is the result of the earth's surface rotating leftward away from the object's path. Because we as observers move with the earth, the object seems to be pushed to the right. This apparent force, called the Coriolis force after its discoverer, acts on all moving objects outside of the equatorial regions; but it is important only in relatively large-scale motions, of the order of the size of a tornado and up. (The spiraling of water down a drain is superficially similar, but the cause is different. It is possible to make the water spiral in either direction, although drain configuration may favor one direction of rotation.)

In the atmosphere more than 1,000 feet above the surface, horizontal air movement is largely parallel to the isobars (lines joining points having the same pressure). At the surface, however, air moving in response to large-scale pressure patterns moves across the isobars with a component toward the center of low pressure. This motion results from the effects of the frictional drag of the surface. The amount of turning toward low pressure is a function of the amount of surface friction. Over rough surfaces,

Figure 1-4. Diurnal variation of wind speed on a mountain top and a nearby lower elevation

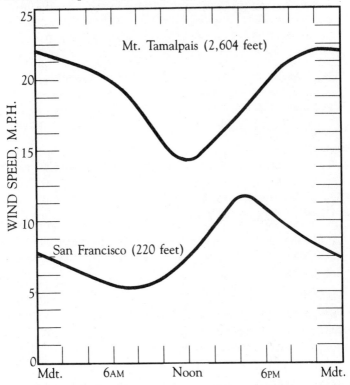

Average March wind speed on Mt. Tamalpais (Marin County) and San Francisco, California. (Source: Monthly Weather Review, Vol. 16, page 513. Sept. 1916)

the angle may be as much as 35 degrees; over the smooth sea surface, the angle between the isobars and the direction of wind flow may be as little as 15 degrees.

Surface friction has another important effect: it retards airflow. Thus, wind speeds measured a few feet above the surface will normally be less than those at about 1,000 feet above the influence of the surface friction. In this friction layer, there is normally a gradient wind speed from near zero within an inch of the ground to the unimpeded gradient wind at the thousand-foot level.

The difference in the wind speed as normally measured by an anemometer a few feet off the ground and that at the top of the friction layer depends on the amount of turbulence in the layer. With a great deal of turbulence and vertical motion, this layer tends to move as a solid body; that is, the top and bottom move at nearly the same speed. This is common in the daytime when unstable conditions prevail. With stable conditions, as on a clear night, the top and bottom are not coupled so tightly; and the bottom is not dragged along so rapidly. Thus, there is typically a diurnal variation in wind speed measured near the surface when there is a marked diurnal variation of stability. It should be noted that this diurnal variation in wind speed may be scarcely evident at the top of a tall building or television tower and may even be reversed at the top of a tall, isolated mountain peak, as indicated in figure 1-4.

If we draw a series of wind vectors (wind arrows with the lengths indicating the speed) representing the wind from the surface to the top of the friction layer, with the bases of the arrows originating from the same point, we will have a graphic representation of the variation of wind speed and direction with height (see figure 1-5).

A line joining the ends of these arrows is the so-called Ekman spiral. Because wind blows counterclockwise around a low (in the Northern Hemisphere), if one stands with one's back to the wind, the region of low pressure will be to the left (Buys Ballot's law). Although this pattern is typical of many situations, especially

Figure 1-5.   Variation of wind speed with height (Ekman spiral)

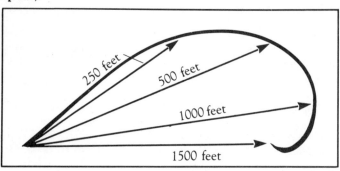

over the sea and flat ground surfaces, topographical and other irregularities may disrupt it completely.

Patterns of Atmospheric Circulations

As indicated previously, unequal heating of the earth's surface leads to vertical motions; these in turn produce high- and low-pressure areas and horizontal motions, which in turn are modified by surface irregularities and the rotation of the earth. From all of these motions, certain patterns may be discovered and described—patterns that will aid in understanding and predicting atmospheric motions. These patterns encompass scales of motion ranging from planetary circulations, such as the trade winds, to the dust devils that swirl across a desert playa.

Chapter 2

Air Masses and
Air Mass Weather

The best hiking and skiing weather of the temperate latitudes occurs when air mass weather predominates. In meteorological parlance, an air mass is a vast mound of air with great horizontal homogeneity and uniformity of properties. Such a mound, covering many thousands of square miles of the earth's surface, can only form when air remains over a uniform surface for a considerable period of time—days or even weeks. If the surface is warm ocean, the air in contact with it becomes warm and moist; if it is frozen arctic tundra, the air becomes cold and dry.

Conditions favorable to air mass formation occur either when air remains for a long time over an area under the influence of a stagnant high-pressure area or when the air has a long trajectory over a uniform surface. In either case, the air assumes the properties imparted to it by the underlying surface. Periodically, air masses migrate from their region of origin and travel many thousands of miles, carrying with them their original characteristics. Thus, a mound of cold arctic air, formed over the frozen tundra, will slide southward, flooding the Great Plains with bitterly cold air often as far south as the Gulf Coast. Eventually, the air becomes modified by its progress over warmer and moister surfaces and loses its identity.

Air masses are characterized by their degree of moistness and

Figure 2-1. Air mass source regions and typical winter storm tracks

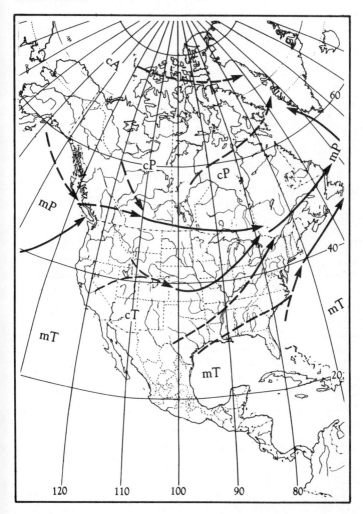

(Source: Reifsnyder, W. E., "What is Weather?" Chapter 1, *Medical Climatology*, ed. by S. Licht, New Haven, CT: E. Licht, Publisher. Reprinted by permission.)

by the area of their origin. If the air mass has developed over the ocean and is relatively moist, it is referred to as a maritime air mass; if it has developed over the continent and is thus relatively dry, it is called a continental air mass. These are symbolized on weather maps by lower-case m and c, respectively. Air masses may also form in tropical regions; these are referred to as tropical (T) air masses. Or they may form in high latitudes, in which case they are referred to as polar (P) or arctic (A) air masses. A curiosity of the naming of these northern air masses is that meteorologists define a polar air mass as one that develops over cold, bare land; an arctic air mass is one that develops over a snow or ice surface. Polar air masses generally develop not over the poles but somewhere near the arctic circle; arctic air masses develop poleward of the polar air masses. The important thing to remember is that arctic air masses are colder and drier than polar air masses.

There are several regions that produce air masses affecting North American weather (figure 2-1). The Bermuda High typically sits over the warm southern regions of the Atlantic Ocean, with its western end pushing inland over the southeastern United States. This produces a maritime tropical (mT) air mass. On the western side of the continent, the tropical air is of Pacific Ocean origin and is caught in the clockwise circulation of the Pacific High. Farther to the north, air over the Pacific is of northern origin and is classified as maritime polar (mP) air. The northern reaches of the continent spawn continental arctic (cA) air masses, and the region just to the south is the source of continental polar (cP) air. South of Greenland lie the cold waters of the North Atlantic, source region for another maritime polar (mP) air mass. The hot, dry air over the deserts of the Southwest forms a continental tropical (cT) air mass. Each of these contributes to the climate of North American hiking regions.

The Bermuda High

Air swirling clockwise around the great tropical anticyclone of the Atlantic Ocean becomes very warm and moist. Especially in the summer, this mass of hot, humid air pushes westward into the Gulf of Mexico and causes a stream of moist air to flow northeastward over the eastern portion of the continent. As it flows over ground heated by the summer sun, vertical currents arise, forming towering cumulus clouds and summer showers and thunderstorms. The humid air may penetrate far north into eastern Canada, bringing periods of oppressive weather. Winds are generally light and variable because the pressure gradients are weak. Local circulations such as the sea breeze and mountain valley circulations (see chapter 4) predominate. Gusty winds occur only in the vicinity of showers and thunderstorms. Visibility is rather poor in the humid, hazy air.

As the air flows northeastward, it comes in contact with progressively less warm ground surfaces. It also becomes somewhat drier as a result of the raining out of the moisture. Accordingly, shower occurrence and precipitation amounts decrease from south to north. For example, in the southern Appalachians, July precipitation is about 6 inches coming in a dozen thunderstorms. Farther north in the White Mountains of New Hampshire, July rainfall is about 5 inches occurring in about eight showers, only five of which are thunderstorms.

The southern portion of the Bermuda High is the spawning ground for the hurricanes that sometimes affect east coast weather. Formed in the band of the easterly trade winds, they generally move westward toward the continent, then curve northward and northeastward around the western margin of the Bermuda High. Most hurricanes affecting the east coast occur in August and September.

In the winter, the Bermuda High retreats eastward. Incursions of maritime tropical air are thus less frequent. Because the ground surface is cooler than the air moving over it, shower activity is reduced; most precipitation comes from frontal storms, as explained in chapter 3.

Pacific Maritime Tropical Air

The west coast of North America is also subject to maritime tropical air originating over the Pacific Ocean. This air, circulating clockwise around the high-pressure area, flows eastward along the northern edge of the Pacific High. Because of its more northerly trajectory, it is somewhat cooler than the mT air on the East Coast and is sometimes classed as maritime polar air. As it approaches the coast, it flows over the cold water brought there by the Japanese current. This cooling from below tends to stabilize the air mass; the clouds that form are stratiform in nature rather than cumuliform, that is, they form layers rather than towering convective clouds. These banks of stratus clouds may stretch from the California coast westward many hundreds or even thousands of miles. The stratus is well known to Californians as the coastal "fog" that blankets the coast for days on end in the summer. The base of the cloud deck may be only a few hundred or a thousand feet above sea level. Frequently, the summer sun is warm enough to dissipate the stratus by midday; but it forms again when the sun sets.

Although this mT air is quite moist, it produces little rainfall except for the drizzle that often settles from the stratus. Curiously, one of the places that may receive substantial precipitation is under the trees. Leaves and needles capture the tiny water droplets as the cloud passes through the tree crowns. Droplets coalesce and grow, finally becoming large enough to run off the leaf or needle. So in the California coastal mountains during a stratus invasion, the driest place is between the trees, not under them.

The stratus is frequently only a few hundred feet thick. Above the layer, the air is clear and relatively dry. Viewing the stratus from above is the spectacular reward for many a hiker in the coastal mountains.

Continental Polar Air

Continental polar air is the delight of the backpacker and skier. Nurtured by long days of gestation over the cool, dry interior of the continent, it appears in the southlands with blue skies and puffy cumulus clouds. During the long nights of the northern winter, snow-covered spruce forest and tundra radiate out much more heat energy than they receive. The air grows cold and a massive high-pressure area develops over the vast subarctic regions. The air swirls around, growing ever colder and denser, but scarcely moister, for the frozen ice and snow evaporate but little. Then some force nudges this mass and starts it sliding southward, flooding the Great Plains. Although shallow, it may move with such inexorable force that it sloshes over both the eastern and western mountains, covering nearly the entire continent with its invigorating crystallinity.

It starts its journey as an extremely stable air mass, with sharp temperature inversions many thousands of feet deep. As it washes over warmer ground to the south the air rises and brilliant white cumulus clouds form in the bright blue sky. There is not much moisture in this air so the clouds remain "cumulus of fair weather" and rain or snow is uncommon. But this onrushing cold air has great power and pushes the warmer, moister air in front of it into great turbulent swirls. Sometimes these swirls develop into giant winter storms; at other times, the triggered storms are smaller (if not less violent) thunder squalls.

Although the weather at the advancing edge of the cold continental air may sometimes be vicious, the weather behind is usually more benign. Skies tend to be clear or partly cloudy, especially some distance into the interior of the air mass. The clouds that do form do not show much vertical development; even when they coalesce into a nearly continuous sheet of stratocumulus, they rarely produce much precipitation. The rain or snow showers that may develop are usually brief. Winds may be gusty, especially near the leading edge of the advancing air mass, and may be moderately strong. Such air masses unfortunately do not linger and usually pass eastward rather rapidly, in perhaps two or three days.

At the onset of a typical continental polar air mass, the wind shifts sharply to the west or northwest as it comes under the influence of the eastern side of the clockwise circulation of air around the central high-pressure area. As the center passes, the winds may become nearly calm, then shift to the south. This usually brings in more moisture and an increased chance of rain or snow. Soon the air mass, modified by its travels over thousands of miles of land surface, passes off and is replaced by another one, perhaps one of maritime origin.

In the summer, the process is similar but less marked. The north country is not frozen but is nevertheless rather cooler than the land to the south. The northern forests and the tundra are now sunlit for long hours; what little moisture there is in the shallow soil is evaporated by the heat of the summer sun. So the cP air starts out rather moister than the winter variety, and it is more prone to develop towering cumulus clouds and thundershowers than its winter counterpart. But most of the time, few shower clouds develop and the air is clear and bright.

For the hiker, this produces some of the finest hiking weather to be found: warm, sunny days and clear, cool nights. But again, the good luck does not last long. In a few days, the wind shifts to the south, bringing in more moisture. Each day the cumulus clouds grow taller until they finally pierce the upper frozen levels with turrets of clouds that erupt with sudden showers and streaks of lightning. In the mountains, the shift of wind to the south as the mound of high pressure moves eastward may be obscured by the complex circulations imposed by the mountains themselves. But the high clouds reveal the process. Watch for an increase of clouds above the cumulus level. This signals an influx of moist air and almost certainly results in increasing shower and thundershower activity in the next day or so. Moisture, as indicated by the upper clouds, is an important clue to predicting a weather change.

Maritime Polar Air of the North Pacific

When continental polar air from the far north takes a track over the ocean, it soaks up both heat and moisture. Although we may think of North Pacific waters as being cold, they are warmer than the frigid cP air from the land. Water is evaporated into the cold air mass and carried upward by the convection triggered by the relatively warm water. Eventually, the continental characteristics of the air mass are replaced by maritime conditions; and a maritime polar air mass is born.

Much of the weather of the west coast of North America is the result of the predominance of mP air. Storms, discussed further in chapter 3, form along the boundary between the cold air to the north and the warm, moist air to the south. Both air masses are moist and the weather produced can only be described as wet. But storms are not even necessary to trigger rainfall from moist mP air. As the prevailing westerly winds push the air over the coastal mountains and the Cascades farther inland, moisture is squeezed out in the form of rain or snow. It sometimes seems the best weather of the west coast is when it is just cloudy and drizzly; the worst is when it pours for three days running.

In the winter, the mP air that invades the north coast is relatively warm and stable as its passes over the colder surface of the continent. It has had a long trajectory around the north side of the subtropical high and is loaded with moisture. Nevertheless, the air tends to be clear and visibility good. But as it slides up the coastal mountains, it quickly cools and condenses into cloud and rain at lower elevations, snow at higher elevations.

In the summer, the subtropical high moves north into the Aleutians, replacing the low-pressure area that persists in the wintertime. The clockwise circulation around the high brings a flow of northwesterly air with a trajectory over relatively cold water. Travel over the warmer land triggers cloudy, showery weather in the cool air. It is a rather monotonous pattern, with frequent showers but few major storms. Clear weather is a rarity, occurring only when the maritime air is displaced by a strong outbreak of continental polar air from far inland.

However, when maritime air from the Pacific flows over the western mountains, it loses much of its moisture thereby becoming warm and dry. The chinook wind of the high plains is much-modified Pacific air.

Maritime Polar Air of the North Atlantic

The North Atlantic is a source region for another maritime air mass that occasionally influences the weather in the northeastern portion of the continent. When mP air from the North Atlantic invades the Maritime provinces and the northeastern states, the weather is cool and cloudy. In the summer, the cold sea surface may produce widespread, low stratus clouds and sea fog (figure 2-2). Most of the time, however, this air mass remains over the ocean and is much more important in affecting Europe's weather than ours.

Desert Air Masses

Over the hot, dry interior of northern Mexico and the southwestern United States, the typical air mass is continental tropical (cT). The surrounding mountains contain the air heated by daily surges of heat from the sun. The highest temperatures in North America are found in this air mass; in Death Valley, the air temperatures reached 134°F on July 10, 1913. But the air is dry,

Figure 2-2.   Stratus

which helps compensate for the high temperatures, at least in terms of human comfort. Death Valley, incidentally, is the only place in the United States where nighttime temperatures often remain above 100°F.

Because of its dryness, the cT air fosters the development of large swings in temperature between day and night. There is little moisture to capture the outgoing thermal radiation from the soil and radiate it back down. Thus, soil surfaces cool off at night, cooling the lower layers of the air as well. The difference between the daytime maximum temperature and the nighttime minimum is often about 40 degrees, whereas in more humid regions, the range is more typically around 20 degrees. Skies are generally clear or nearly cloudless, for the subsiding air characteristic of this region tends to dissipate rather than form clouds. This is also the reason for the low humidities, often as low as 10 percent: the ultimate source of the air is high above the surface, far from sources of moisture.

Chapter 3

Storms and Frontal Systems

A front is the battleground where two air masses clash. Whereas the weather associated with an air mass is usually rather benign, that associated with frontal storms is often wet, windy, and wild.

An air mass is a shallow mound of air with great horizontal homogeneity in temperature and humidity. Nearby air masses may have different characteristics. Sometimes the gradation from one air mass to another is rather gradual: on a weather map, it may be difficult to determine where one stops and the other starts. Usually the boundary is well marked by sharp discontinuities of temperature, humidity, and wind speed.

But fronts are far more than just boundaries between two air masses of different characteristics. They are regions where the great weather producers—cyclonic storms—originate. The sharp contrasts of temperature and wind found along a front frequently generate low-pressure areas with counterclockwise winds (in the Northern Hemisphere) swirling in toward the center. As the air moves inward, it replaces air being sucked upward and outward from the center of the low. This upward motion produces clouds and rain as the air cools and ultimately reaches saturation.

Thus, a cyclonic storm (the term "cyclonic" refers to the counterclockwise circulation around a low-pressure center) is born at a frontal zone, and its development and eventual decay are inextricably bound up with frontal structure. To understand this process, we must trace the three-dimensional history of the events that occur along the front.

One might suppose there is a continuous circumpolar front marking the boundary between polar air to the north and tropical air to the south. Indeed, there is such a frontal zone, the so-called polar front, that stretches around the polar regions in a more or less continuous band. It is along this polar front that our mid-latitude storms form. To envision the structure of the polar front, one might imagine the polar regions covered with a layer of thick molasses, oozing down over the temperate latitudes. Such a layer would have a sharp leading edge, nearly vertical at its forward edge but gradually sloping northward at upper levels. As the polar front pushes southward, it displaces the moist tropical air ahead of it. Some of the tropical air may be forced to slide up over the advancing cold air and may be evident above the polar air as high clouds; but typically, the cold air completely displaces the tropical air and the skies in the polar air are mostly cloudless.

Just behind the advancing polar front, the winds will typically be from a northerly direction as the air swirls clockwise out of the high-pressure area associated with it. To the south, winds in the tropical air will have a southerly component, as, for example, Gulf air moves northward along the western edge of the Bermuda High. Periodically, these two countercurrents initiate the process of "cyclogenesis" or storm formation along the polar front: a dimple forms along the boundary.

Sometimes this dimple travels eastward along the polar front, much like a wave travels along a jiggled rope. A small low-pressure area may be associated with the dimple, causing clouds and brief periods of precipitation as the moist air from the south is forced upward over the cold polar air. Such wavelets usually move rather rapidly along the polar front and eventually die out.

At other times, dimples on the polar front expand rapidly into giant waves that impose a completely different circulation and weather pattern on the front. They develop into giant storms that affect the weather for many thousands of square miles. The winds and weather that accompany these storms form regular patterns that enable us, by observing them closely, to predict the weather many hours ahead. By observing the progression of weather events from a single location, we are often able to tell where we are with respect to the storm system and how the system

Figure 3-1.   Series of storms on the polar front

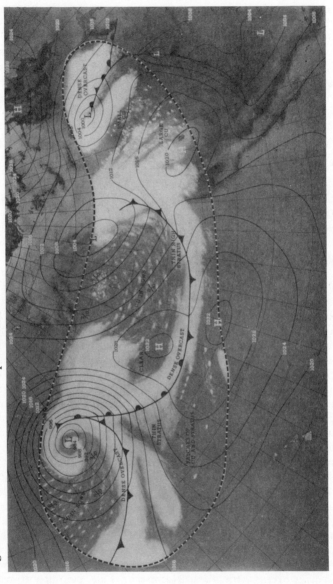

is moving. From these local observations and from a knowledge of the regularities of storm movement, we are in a good position to make useful forecasts of the sequence of weather events.

To do this, we should first see how such a developing storm would appear from the vantage point of an astronaut high above the surface whose gaze could encompass the entire storm system. Figure 3-1 is an actual photomosaic of such a storm system together with a pictorialized weather map showing the associated pressure patterns and frontal systems. The three low-pressure areas show storms in various stages of development, all rather well along toward eventual decay and dissipation. So our story of storm development must start at an earlier stage.

As indicated previously, storms develop on the polar front in regions where the winds on opposite sides of the front and from opposite directions cause a shearing of the wind across the front. Such a region might be the region between the two high-pressure areas near the center of the photo map. North of the polar front, the wind spiraling clockwise around the high would be from the northeast. South of the front, the wind around the high-pressure area found there would be from the southwest. If other conditions necessary for cyclogenesis were present, a low-pressure area would form along the front and a closed counterclockwise circulation would develop.

On the eastern side of the low, the warm, moist air from the south is forced northward up and over the dome of cold air north of the front (figure 3-2). As the air is forced upward, it cools and eventually reaches saturation. At this level, clouds form, developing a shield that may stretch 1,000 miles ahead of the front. At the same time, the bubble of cold air is itself forced northward; the front as depicted on the weather map also moves northward. Because the front is moving in the direction in which the warm air is moving, it is referred to as a warm front.

On the west side of the low, cold air is being pushed southward by the counterclockwise circulation. It pushes the surface front along and works its way under the warmer air ahead of it. This cold front—so called because the front is at the leading edge of the advancing cold air—also triggers cloud formation. Because of the steepness of the cold front and the instability caused by cold air moving over a warmer surface, the clouds are likely to be towering cumulus clouds that produce showers and thundershowers. Well behind the front, the air is drier and can support only puffy cumulus clouds with little or no precipitation.

Figure 3-2. Three-dimensional representation of developing frontal storms

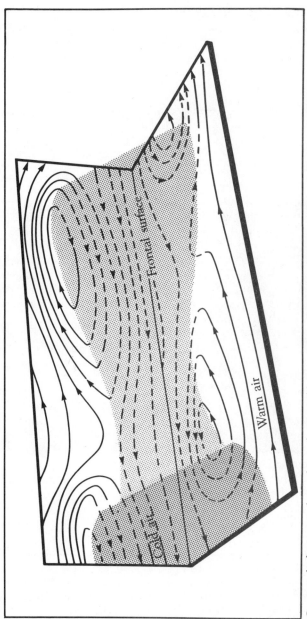

Counterclockwise wind circulation brings cold air southward, forcing warm, moist air aloft. (From NCAR Quarterly, Issue No. 11, July 1965. Reprinted by permission.)

Figure 3-3. Frontal storm
(Base photograph, courtesy U. S. Dept. of Commerce, National Oceanic and Atmospheric Administration. Cross-sections added.)

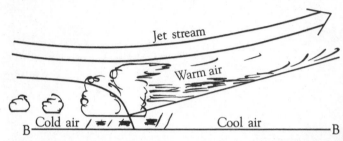

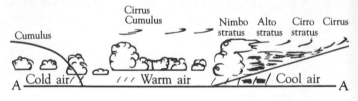

A typical sequence of clouds and precipitation is shown in the vertical cross section A-A (figure 3-3). This cross section is taken along the general direction of movement of the storm system. At the eastern end, in the high-pressure area, few clouds are

Figure 3-4. Cirrus

Figure 3-5. Cirrostratus

present, for the downward sinking of air in the high tends to dissipate clouds. A little bit to the west, high cirrus clouds are present (figure 3-4), representing the farthest reach of the over-running warm, moist tropical air. Cirrus clouds are distinguished by their indistinct, fuzzy edges and are composed of ice crystals. The "mare's tails" shown in the photograph are harbingers of

Figure 3-6. Altostratus

precipitation if they become thicker and lower and if the surface wind is from the east or south.

Closer to the surface front, the overrunning warm air is moister and the clouds are thicker. A veil of cirrostratus covers the sky. The ice crystals of the cirrostratus often produce a halo around sun or moon (figure 3-5). Closer to the surface front, the clouds are no longer high enough to be composed of ice crystals but are now made up of water droplets. These altostratus clouds (figure 3-6) are thick enough to obscure the sun. They generally occupy levels between 10,000 and 20,000 feet above the surface.

As the front is approached, rain (or snow, if temperatures are low enough) develops. The dominant cloud visible from the ground is nimbostratus (figure 3-7), ragged patches of rain cloud obscuring the thick layers of altostratus and cirrostratus above. The surface front itself may not be very well defined, although the rain shield usually terminates close to the front. The wind may veer (turn clockwise) to the southwest and the pressure, which drops as one approaches the front from the east or northeast, levels off.

The area between the cold and warm fronts is filled with warm, moist mT air and is appropriately called the warm sector. It is the temperature-contrasts that are important, not the actual temperatures. Thus, in the winter, warm sector air may be quite

Figure 3-7.   Nimbostratus

cold, even though it is warmer than the air on the other side of the front. Clouds may be present in the warm sector; if so, they are typically cumulus clouds with flat bases and puffy tops or stratocumulus formed by the coalescence of individual cumulus clouds into a continuous layer (figure 3-8). Afternoon showers may be triggered by daytime heating; nighttime showers are rare. Middle-level clouds such as altocumulus (figure 3-9) and higher cirrus may also be present.

The cold front forming the western and northwestern boundary of the warm sector is frequently marked by a line of thunderstorms triggered by the sharp upward thrust of the warm air by the advancing cold air. Sometimes ahead of the actual front there is a precursor line of thunderstorms along a squall line. The clouds here are towering cumulonimbus, discussed in chapter 5.

Behind the cold front lies a region of clear, dry air. Although clouds may be present for some distance behind the front, the dry air cannot support extensive cloud formation. The flat-bottomed cumulus clouds that do develop are scattered and not very tall although occasionally they spread out to form layers of stratocumulus. Because of the kink in the isobars (lines of equal

Figure 3-8. Stratocumulus

Figure 3-9.   Altocumulus

pressure on the weather maps), the wind shifts clockwise as one passes through the cold front. Pressure rises as one gets farther away from the center of low pressure. Farther behind, a new mound of high pressure is encountered and the cycle is complete.

As time goes on, the wave deepens and expands; that is, the central pressure of the low falls and the area of closed circulation expands. The cold front speeds up and eventually catches up with the warm front that is receding ahead of it. This process of closing or occluding the wave eventually eliminates the warm sector entirely. All that is left is the remnant of a front stretching southward from the center of low pressure. This so-called occluded front may be the remains of either the original warm front or the cold front, depending on the relative temperatures of the air in the three regions of a wave. If the air behind the cold front is colder than the air ahead of the warm front (both are in the cold air mass), the advancing cold frontal surface actually lifts the warm front and the air ahead of it. This is a cold-front type occlusion; the weather associated with it is likely to be that normally associated with a cold front: squally, thundery weather as the front passes. If the temperatures are reversed, the advancing cold front slides up over the retreating warm front and a warm-front type occlusion develops. The associated weather is similar to warm-front weather.

The occluding wave in central Canada (figure 3-1) is shown in more detail in the cross section B-B on figure 3-3. It is a cold-front type occlusion, as shown in the accompanying cross section, except that there is no warm sector and the cold-front clouds appear near the occluded front.

Of particular interest here is the location of the jet stream, the band of high-speed air present high in the atmosphere. Winds in the jet stream may reach several hundred miles per hour in its west-to-east sweep around the North Pole. (The jet stream was discovered by high-flying but slow World War II bombers flying westward across the Pacific. When unknowingly caught in the jet stream—there was no visible boundary—their westward progress was slowed almost to zero. There were times when the planes actually moved backward relative to the ground.)

The final stages of a low-pressure storm occur when the low-pressure area fills and the occluded front dissipates. The polar front reforms, perhaps by a subsequent push of cold polar air from the north, and the stage is set for another wave development.

Because waves travel in a general west-to-east direction along the polar front, the sequences of clouds and weather described along the cross sections occur in the same order to the observer on the ground. By watching the sequence of clouds above and the shifts in wind near the surface, careful observers can locate themselves with respect to the system. If the system is well behaved (unfortunately, not all are), accurate forecasts can be made of the weather for the ensuing twelve to forty-eight hours. It usually takes between twenty-four and forty-eight hours for a major weather system to pass a fixed point on the ground.

Table 3-1 summarizes these changes. All the weather elements except atmospheric pressure can be either observed visually or sensed by the alert recreationist. Even temperature and humidity changes can be determined without instruments, although a pocket thermometer is a useful device to carry. A pocket altimeter, although relatively expensive, is a most useful device for the amateur weather forecaster. Most recreationist's altimeters also contain a pressure scale, marked in millibars or inches of mercury. As long as the altimeter is kept at the same elevation (as when camping for the night), pressure changes can be monitored for rapid rise or fall.

It is not difficult to develop skill in forecasting the weather from a single location. But it does require a systematic approach to observing the weather and practice at interpreting the observations. The following procedures may prove useful:

1. Practice identifying cloud types and interpreting their significance. Note their direction of movement and their sequence. If they are cloud types or sequences usually associated with frontal storms, try to place yourself in the sequence; that is, try to locate your position on the pictures and cross sections of frontal storms presented in this chapter.

2. Make systematic visual or instrumental observations at approximately the same times each day. Keep a simple log or diary of these observations. Making these observations at the same time permits identification of day-to-day trends. The variation of temperature throughout the day that tends to obscure the day-to-day trend is thus eliminated. Are there high clouds present today when there were none yesterday, signifying the approach of a moist air mass at an upper level? Is the wind stronger today than at the same time yesterday, indicating the approach of a more vigorous circulation pattern? Are daytime cumulus clouds appearing earlier and building up higher each day, leading to a greater chance of afternoon showers developing? Or are the clouds dissipating earlier, indicating a turn to drier air?

3. Observe the surface wind speed and direction. Note especially marked changes. Are the winds of local origin (see chapter 4), or are they driven by an approaching large-scale weather system?

4. Make a forecast each day of what you expect the weather to be in twelve and twenty-four hours. Write it down in your diary or log and check it the next day. Analyze why your forecast was correct or why it went wrong. This systematic feedback will rapidly improve your forecasting skill.

5. Keep your eye on the sky. With a heavy pack and a steep trail, it may be difficult to look at anything but the trail 5 feet ahead. But whenever you stop for a breather, look up. Not only is the ever-changing panoply of clouds a thing of beauty in its own right, it has something to tell us. The more we look and interpret, the more meaningful the sky changes become. Our appreciation of the natural world will be more complete for having added the atmosphere to our catalog of natural beauties.

Table 3-1. Typical Weather Conditions Associated with Approach and Passage of Fronts and Storm Systems

Phenomenon	Warm Front		Cold
	Approach	Passage	Approach
Pressure	Falls steadily	Levels off, or falls unsteadily	Falls slowly; or rapidly if storm intensifying
Wind	SE quadrant; speed increases	Veers to S quadrant	S quadrant; may be squally at times
Clouds	Cirrus; cirrostratus; altostratus; nimbostratus; thickening	Stratocumulus; sometimes cumulonimbus; clearing trend	Cumulus or altocumulus; cumulonimbus in squall line
Precipitation	Steady rain or snow starts as clouds thicken; intensifies as front approaches	Precipitation tapers off; may be showery	None or showery; intense showers or hail in pre-front squall line
Temperature	Increases slowly	Slight rise	Little change or slow rise
Humidity	Increases	Increases; may level off	Steady; or slight increase
Visibility	Becomes poorer	Becomes better	Fair; may become poor in squalls

Front	Occluded Front		Observer North of Frontal System
Passage	Approach	Passage	
Sharp rise	Falls steadily	Rises, often not as sharply as cold front	Falls slowly; then rises slowly as system passes
Sharp veer to SW quadrant; speed increases; gusty	E quadrant; may veer slowly to SE quadrant; speed increases	Veers to SW quadrant; speed decreases	NE quadrant; backs through N to NW quadrant
Cumulonimbus; sometimes few clouds; clearing trend	Cirrus; cirrostratus; altostratus; nimbostratus	Slow clearing; stratocumulus; altocumulus	Cirrus; cirrostratus; altostratus; nimbostratus; stratocumulus; cumulus
Showery; perhaps thunderstorms; rapid clearing	Steady rain or snow starts as clouds thicken; intensifies as front approaches	Precipitation tapers off slowly	Rain or snow starting as clouds thicken and lower; slow clearing
Sharp drop	Slow rise	Slow fall	Steady or slow decrease
Sharp drop	Slow increase	Slow decrease	Increase; slow decrease as storm passes
Sharp rise; becomes excellent	Becomes poorer	Becomes better	Becomes poor; slow betterment

Chapter 4

Microclimate and Mountain Meteorology

When the skies are stormy, the weather is everywhere nearly the same—even across a frontal storm system covering many thousands of square miles. When skies are clear, local conditions may vary markedly. At noon on a sunny day, a south-facing slope may be 10 or more degrees warmer than a north-facing slope only a few hundred yards away. On a clear, calm night, the temperature in a small topographic depression may be 20 degrees colder than on a nearby slope.

The climate of these highly localized places is called microclimate. The backpacker can use these localized climates in selecting a place to pitch a tent or stop for lunch. Such knowledge may spell the difference between a pleasant night's sleep and a wakeful night of shivering.

Microclimates are evident only during certain types of weather. They develop most characteristically during the periods between storms: the so-called air-mass weather when skies are clear and winds are light. During storms, the factors that produce marked microclimate differentiation are absent. What are these factors? How do they operate to produce microclimates? To answer these questions and to alert readers to backcountry conditions conducive to favorable or unfavorable microclimates, I shall discuss the environmental factors that are important in determining microclimatic differentiation.

Heat Transfer and the Energy Budget

Radiation. The energy that heats the earth and drives the atmosphere comes from the sun. Without the sun, the earth would be a very cold place, indeed, a dead planet. The radiation from the sun, passing through space and the earth's atmosphere at the speed of light, is absorbed by the air, the ground, the water in the sea, and every other thing it hits and is transformed into heat energy. It is this transformed energy we feel when we turn our face to the bright sun on even a cold winter day. This absorbed energy is the fuel that warms everything it touches.

But not every material absorbs solar energy equally well. A totally black surface absorbs nearly all the sun's visible rays; a white or shiny surface absorbs very little. However, about half the sun's radiation is composed of invisible, near-infrared radiation; and the visual darkness of a surface does not necessarily indicate its absorptivity to these longer waves. (Note that dark leaves photographed with film sensitive to the near-infrared rays may appear quite light, indicating little absorption and high reflection of this portion of the solar spectrum.) As a general rule, dark, rough surfaces absorb more of the sun's rays than light, smooth ones. This means that, other things being equal (they usually aren't), dark, rough surfaces will be warmer than smooth, light ones exposed to the same solar radiation.

The major control over the amount of radiation a surface receives (as opposed to absorbs) is its orientation with respect to the sun. A surface perpendicular to the sun's rays would, in the absence of the atmosphere, receive nearly 2 calories (small calories—a food calorie is 1,000 times larger) per square centimeter each minute. If the sun's rays hit the surface at a 45-degree angle, the amount received per square centimeter is 70 percent of that value. At 30 degrees, the value is 50 percent; at 20 degrees, it is about 33 percent; and at 10 degrees, it is less than 20 percent of the perpendicular amount. (This follows from the so-called cosine law of illumination, which states that if a surface perpendicular to the rays receives one unit of radiation, a surface tilted at some angle from the perpendicular will receive radiation equal to the cosine of that angle.) The atmosphere modifies this relationship somewhat. Nevertheless, an east-facing vertical cliff may receive more radiation from the sun shortly after sunrise than a horizontal surface receives at noon. This depends, of course, on the latitude and time of year.

The amount of solar radiation the ground receives is also dependent on the clarity of the atmosphere and the elevation at the point of measurement, which determines the thickness of atmosphere between that level and the upper limit of air. The turbidity of the atmosphere determines the partition of the sun's radiant energy between the direct beam of the sun and the scattered sky radiation. On a very clear day at high elevations, nearly 90 percent may come directly from the sun, the remainder from the sky. On a very hazy day near sea level, the split may be more nearly fifty-fifty with half the solar energy coming directly from the sun and half from the sky.

Solar radiation is not the whole story, however. Once the sun's energy is absorbed and turned into heat, the warmed object radiates energy by virtue of its temperature. At normal terrestrial temperatures, this radiation is in the form of thermal radiation in the very far infrared region of the spectrum. It is not visible and goes on day and night, as long as the radiating object has a temperature above absolute zero. Indeed, because the amount of radiation depends on the temperature, a hot ground surface in the daytime emits more thermal radiation than the cooler ground surface at night. The sky also radiates energy in the far-infrared region from particulates, water droplets, and various molecules such as carbon dioxide and water (figure 4-1).

Day or night, there are various types of radiant energy streaming up and down at the earth's surface. The net amount of radiant energy the surface gains or loses can be calculated with a simple bookkeeping procedure. On a sunny summer day, the surface gains radiant energy because of the dominant influence of the sun. On a clear night, however, the surface loses more radiant energy than it receives from the sky because the sky is colder than the surface. The net radiation is outward. The result is surface cooling through the course of the night. If a layer of low cloud spreads over an area, the stream of down-coming radiation from the warm cloud may equal or even exceed the upward stream from the ground surface. If there is then a net radiation downward, the ground surface will start to warm up rather than cool down, as is typical at night. (The infrared radiation from the ground is not reflected from the cloud, but absorbed and reradiated. The distinction is important, for the cloud does not act as a passive reflector; it radiates at its own temperature. A high, cold cloud

Figure 4-1. Radiation streams at the earth's surface

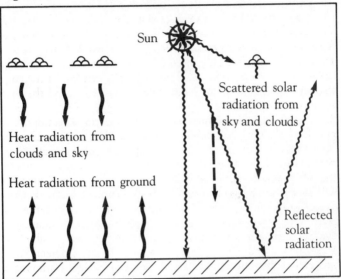

will radiate downward less energy than a low, warm cloud. If reflection were the process, the temperature of the cloud would make no difference.)

Thus, the net radiation can be considered the surplus or deficit of heat energy at the surface of the ground (or any other surface or object, for that matter). When the net is positive, the surplus heats the ground surface and the air in contact with it and may evaporate water. When the net is negative, common on a clear night, heat must be supplied from the ground or air or by condensation of water vapor in order to warm the surface, etc.

Ground heat conduction. A surplus of radiant energy absorbed by the ground surface is immediately turned into heat and is perceived as surface temperature. If the surface is hotter than the ground beneath it, heat is conducted downward, warming the lower layers. (*Conduction* is the way heat is transferred through an opaque solid.) The thermal properties of the soil determine how and how much of the heat is stored. If the surface layer is a good heat insulator (perhaps a thick layer of dry leaf and needle litter), little heat can be conducted downward, though the temperature of a surface exposed to the sun may become very high. Most of this heat will be transferred to the air.

On the other hand, a good conductor with a large heat capacity, a granite rock, for example, will conduct heat downward. The surface will not become as hot as the leaf litter. (It may feel hotter to the bare foot because the stone is capable of transferring more heat to the skin in contact with it even though the initial surface temperature is lower. This can easily be checked by placing a small thermometer on the two surfaces to compare their temperatures.)

If one's nighttime objective is to keep warm, a good place to put one's sleeping bag is on a large rock that has been heated thoroughly by the sun. It will continue to supply heat for a long time during the night. A thick litter layer is next best; even though it may be cooler than the rock, its insulating properties will slow the heat loss from the body downward. The poorest choice would be a cold and perhaps damp soil that would conduct body heat downward. Water is a good conductor of heat and damp soils usually conduct heat better than dry soils.

Heat convection to and from the surface.   As the surface is heated by the sun, its temperature rises, causing heat to be driven into the ground by conduction. But the air layers in contact with the ground are also heated. This warmed air becomes lighter and starts to rise because of its bouyancy, carrying its heat upward. This process, called *thermal convection,* is another way in which heat is transferred to the air from a heated surface. When the air is cooled by contact with cold ground, heat is transferred downward to the surface; but the process is slower and less efficient than the upward convection.

A wind blowing on a surface also carries heat to or from a surface, depending on whether the air is warmer or colder than the surface. This *forced convection* depends on the wind speed and is important, for example, in cooling the body when a cold wind blows on it. This windchill phenomenon is discussed in greater detail in chapter 5. I shall note here only that forced convection is generally more effective in transferring heat from a surface than is thermal convection: even a slight wind on a cold day produces a markedly greater cooling effect than is present in a dead calm.

In natural conditions, the two types of convection almost always operate together. In calm, sunny weather, thermal convection may dominate the convective heat transfer process; in windy weather, forced convection will predominate.

Evaporation. The last way heat is lost from a surface is by *evaporation*. Heat is required to evaporate water—540 calories are needed to evaporate 1 gram of water at room temperature—and the result is marked surface cooling. Whether evaporation can occur depends, of course, on whether water is available. Dry, bare ground or a rock surface has little water available, and evaporation is minimal. A well-watered meadow or forest has a plentiful supply of moisture; evaporation may proceed apace. Indeed, evaporation from such a surface may use up most of the radiant energy available, leaving little for heating the ground or air. This is one reason, for example, air over the ocean shows little temperature change from day to night and air temperatures over a desert fluctuate more diurnally than over a forest.

Though evaporation produces surface cooling, the transition of water from liquid to vapor occurs isothermally, that is, at a constant temperature. Thus, in a boiling teakettle, water at 100°C is changed to steam at 100°C. However, a great deal of heat energy is required for this transition, about 600 calories per gram of water boiled away. This heat energy is stored in the water vapor as latent heat and released when the water vapor condenses back to liquid water. When liquid water evaporates at lower temperatures, nearly as much heat is required. If the heat comes from a wetted surface (for example, skin), the temperature of that surface will go down. Evaporation thus results in cooling the evaporating surface.

There are other sources of the heat energy available for evaporation, however. If the evaporating surface is exposed to sunlight, for example, the heat may come from the radiant energy, and the surface may not change its temperature. A similar process occurs when the teakettle boils: the flame under the kettle supplies heat. As the water boils away, utilizing this heat, the temperature of the boiling water remains the same.

When water changes from solid to liquid, a similar process occurs. One gram of ice requires 80 calories to melt; the heat is stored as latent heat, which is released when the reverse process of freezing occurs. It is also possible for ice to go directly to vapor. The process is called sublimation and requires 677 calories to change 1 gram of ice to vapor at 0°C.

The processes of evaporation, condensation, melting, and freezing require prodigious amounts of heat as compared with normal heating and cooling. Whereas only 1 calorie is required to increase the temperature of 1 gram of water 1 Celsius degree and

100 calories are required to heat 1 gram of liquid water from 0°C to 100°C, nearly 600 calories are required to evaporate that 1 gram of water. So it is not surprising that these processes exert a strong stabilizing effect on temperatures at the earth's surface. Cross-country skiers are aware that snow temperatures may fluctuate greatly below freezing. But once melting starts, the snow surface temperature remains at 0°C no matter now high the air temperature or how wet the snow.

Heat balance.   The various flows of heat energy to and from a surface do not occur randomly and independently but operate together in an organized, predictable way (figure 4-2). Because energy can neither be created nor destroyed, all the flows must balance, in much the way flows of money into and out of a bank must balance. Deposits and withdrawals may represent the radiative and convective flows of energy. A net gain of funds in the

Figure 4-2.   Heat balance at the earth's surface

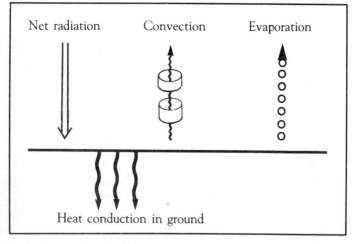

bank finds its way to the vault, where it is stored, only to be withdrawn when withdrawals exceed deposits. (One may hope evaporation is not one of the processes operating in this analogy.)

In the natural *heat balance*, the surface temperature is the variable that acts to keep the system in balance. If the sun comes from behind a cloud and the surface absorbs more radiant energy, the surface temperature increases. This causes more heat to flow

Figure 4-3. Diurnal cycle of heat balance over a bare ground surface

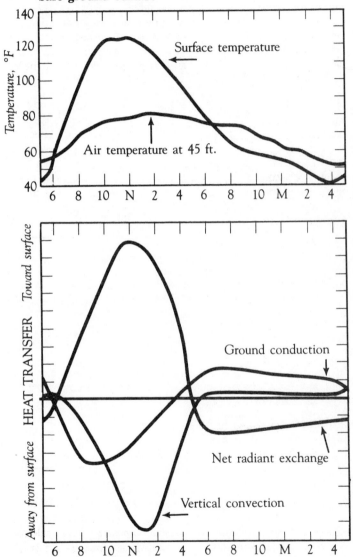

(Adapted from Brooks, F. A., *An Introduction to Physical Microclimatology*. Davis, CA: University of California, 1949)

into the ground by conduction and more heat to flow to the air by convection. The heated surface will radiate more heat outward, also; and these outward flows will exactly balance the increase in absorbed solar radiation. If the surface is wet, evaporative heat loss must also be added to the other losses to keep the flows in balance.

An example of a typical diurnal progression of the heat balance at a dry soil surface on a sunny day is snown in figure 4-3. In this diagram, all the radiant energy flows are combined into the net radiation. Before sunrise, the net radiation is outward, representing a heat loss to the cold night sky. This must be balanced by a heat flow upward from the ground. Little convective heat transfer occurs at this time. The result is a gradual decrease in the surface temperature and in the air above it. As soon as the sun rises, the net flow of radiant energy is toward the surface, and a large surplus develops. This causes the surface temperature to rise, and heat is driven into the ground by conduction and into the air by convection. As a result, the air temperature increases. Late in the afternoon, the net radiation again becomes outward as the sun sinks low in the sky and ceases to overcome the thermal radiation emitted by the hot ground surface. Thus, the surface cools and the air in contact with it cools also. This cooling process continues throughout the night until the sun rises once more to start the cycle over again.

If the surface were wet, evaporation would also produce a heat loss to it. The major effect would be that the surface temperature, and thus also the air temperature, would not rise as much in the daytime as it does over dry ground. That is, moisture has an ameliorating effect, keeping the diurnal temperature range lower than it would be over dry ground.

Atmospheric Stability and Thermal Convection

The atmosphere is in constant motion: watch the smoke from a chimney or a campfire on even the calmest day. Some drifting will always be observed. "Calm" in weather service terminology is somewhat arbitrary. Although defined as those periods when the wind is blowing at less than 1 mph, it really depends on the sensitivity of the wind-measuring equipment. Most standard anemometers will not respond to winds less than about 2 mph.

We are generally aware of the horizontal wind, that measured by an anemometer and reported as the surface wind; we are much less aware of the vertical motions of the atmosphere. Vertical currents are always present, although usually of a smaller magnitude than the horizontal currents. The exception to this is in the towering cumulus clouds that produce showers and lightning and may contain upward-moving air currents exceeding 60 mph. Nearly all precipitation results from vertical motions in the atmosphere; and the key to understanding these vertical motions lies in the role atmospheric stability plays in producing or diminishing them.

Consider for a moment a small rubber balloon carried by a child on the way up to the observation platform on top of the World Trade Center in New York City. Two things will happen as the high-speed elevator carries the balloon upward: it will expand slightly, increasing its volume about 4 percent. And if we had a thermometer inside the balloon, we would notice that the temperature decreased about 7 degrees Fahrenheit.

Because the process was carried out rather rapidly, in the time it took for the elevator to go the distance, little heat was gained or lost through the skin of the balloon. Thus the process can be considered to be "adiabatic," that is, without exchange of heat between the air in the balloon and that outside. If we brought the balloon back down to street level, we would find that the original volume and temperature were regained, proving that no heat had been gained or lost in the total process. The rate at which temperature changes as the pressure changes is known as the *adiabatic lapse rate*. Since pressure in the atmosphere decreases upward at a known rate (figure 1-1), the adiabatic lapse rate can also be expressed as a rate of change of temperature with change in altitude as a parcel of air is transported from one level to another. At sea level, this rate of change is approximately one Celsius degree per hundred meters, or about 5½ Fahrenheit degrees for each thousand feet of elevation difference.

Many vertical motions in the atmosphere can be considered to be adiabatic. Air pushed over a mountain range by a large pressure system will cool at very nearly the adiabatic lapse rate. Suppose that the air that starts out at the bottom of the windward slope of a mountain has a temperature of 68°F (figure 4-4). As it is pushed up the slope, it decreases 5½ degrees for every thousand feet of elevation gained. At the top of the 6,000-foot ridge, it will have cooled adiabatically to 35°F. On the way down the other side, the air will heat adiabatically until it reaches the bottom at the same temperature it had when it started, 68°F. The air very close to the surface may be heated or cooled by contact with hot or cold ground, and the picture may be modified slightly. However, air a hundred feet or so above the ground surface will follow the adiabatic rate closely.

Figure 4-4. Air temperature changes as air moves over a mountain range

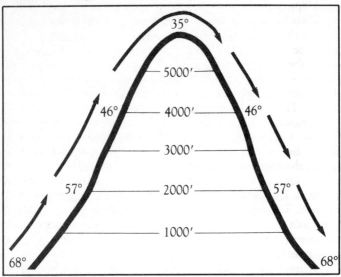

What happens if the air becomes saturated with moisture as its ascends the mountain slope? To answer this question, we must understand the role of water in the atmosphere. For every possible combination of atmospheric pressure and temperature, there is a maximum amount of water vapor the air can hold. If there is a source of liquid water available and if the air and water are at the same temperature, evaporation will continue until the air is fully loaded with moisture. Evaporation will cease and the moisture content of the air will remain constant at the fully saturated level. (More precisely, net evaporation will be zero; evaporation and condensation, the return of water vapor to the liquid surface, will be in balance.) The relative humidity of the air, the ratio of the amount of water vapor in the air to the maximum amount the air can hold, is 100 percent. For example, at sea level, air at a temperature of 86°F will contain 26 grams of water vapor in each kilogram of air that is saturated. At 68°, the maximum amount is 14 grams per kilogram; and at 32°, it is 4 grams per kilogram.

Air near the surface is rarely saturated, that is, at 100 percent relative humidity. When it is at 100 percent, the air feels very moist indeed, for perspired water cannot readily evaporate and the skin becomes moist. We feel either cold and clammy or hot and drippy, depending on the temperature.

What has this to do with adiabatic lapse rates? To see, look again at the air flowing over the mountain, but this time with a difference. Suppose there is some moisture in the air that starts its way up the windward slope. As long as the air is not saturated, it will cool very nearly at the adiabatic lapse rate. But further lifting and cooling beyond the point of saturation will result in supersaturation. This cannot happen in the atmosphere (except to a very slight degree); so some of the water vapor will condense into tiny droplets, forming fog or clouds. However, because evaporation requires heat energy (nearly 600 calories to evaporate 1 gram of water), the condensing water must release this heat energy to the air. The result is that the further cooling of the air as it is forced to ascend is less than it was before saturation and condensation occurred. This cooling rate is about 60 percent of the unsaturated rate, about 3 Fahrenheit degrees per thousand feet. This rate is called the *saturated* or *moist* adiabatic rate, to distinguish it from the *unsaturated* or *dry* adiabatic rate.

Figure 4-5 shows what will happen to the ascending air as it moves up the mountain slope when it reaches saturation. As long as the air is unsaturated, it cools at the dry adiabatic rate; in our example, it cools at this rate up to 4,000 feet. At this point, the air reaches 100 percent relative humidity; further lifting results in cooling at the moist adiabatic rate and the temperature at the top of the mountain is 40° (compare figure 4-4). Cloud forms at the 4,000-foot level and the ridge is enshrouded. The air in the cloud is at 100 percent relative humidity; the excess water exists as liquid droplets. As the foggy air slides down the leeward slope, it starts to warm at the dry adiabatic rate (as soon as its temperature increases a little bit, the air can hold more water vapor and the relative humidity therefore decreases). But because there is a ready supply of liquid water available, evaporation occurs, which slows down the rate of heating (heat energy is required for evaporation). The net result is that the downward-moving air reheats at the moist adiabatic rate until all the cloud droplets are evaporated. This will occur at the 4,000-foot level on the leeward slope, and the temperatures will be symmetrical on the two sides of the mountain.

This is the process that forms the lenticular clouds (Altocumulus lenticularis) often seen above mountain ridges, particularly when the jet stream is present high above the ridge. Air is forced upward until it reaches saturation, when it becomes visible as a cloud. As the current of air reaches its maximum height and starts moving downward, evaporation occurs but is not complete until the current reaches the same level it had when condensation first occurred. The result is a lens-shaped cloud that is concave downward. Because of variations in the amount of moisture contained in various air layers, there may be several separated lens-shaped segments piled one on top the other (figure

Figure 4-5. Air temperature changes as air moves over a mountain with cloud formation

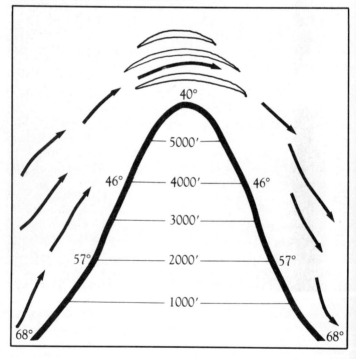

Figure 4-6.   Altocumulus lenticularis

(U.S. Dept. of Commerce, National Oceanic and Atmospheric Administration)

4-6). It is apparent that the cloud itself is not moving, even though the air flowing through it may be traveling at very high speeds, often in excess of 100 mph. The undulations in the air currents imposed by a massive mountain range such as the Sierra Nevada of California may produce spectacular cross-wind rows of lenticular clouds parallel to the range and diminishing in size downward. This is the well-known Sierra wave familiar to hikers in the eastern portion of the range near Owens Valley.

To return to the example of air flowing over a mountain, what happens if some of the condensed water falls out as rain or snow? Now some of the water is physically removed from the air current and is no longer available for reevaporation into it on the lee side. As a result, the heat added to it in the condensation process remains in the current as it slides over the ridge top. Heating on the lee slope starts out at the moist adiabatic rate.

Figure 4-7. Air temperature changes as air moves over a mountain with precipitation

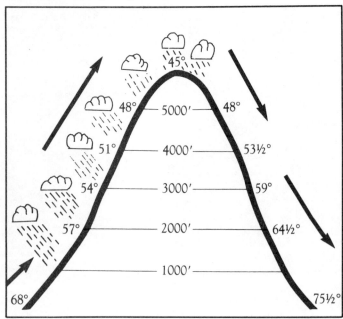

After the clouds evaporate, further heating occurs at the dry adiabatic rate and the air reaches the bottom of the lee side warmer than when it started out (figure 4-7). In the example shown, the air that started out at 68° reaches the same level on the lee side at 75½°. Visitors to the slopes and valleys on the lee side of any major mountain range are familiar with this phenomenon. The generic name is foehn (pronounced "fern"), but we call them chinooks, Mono winds, East winds, Santa Anas, and many other names, depending on the locality.

The air in a chinook or foehn is typically exceedingly dry, with relative humidities often below ten percent. It is also relatively hot, having been warmed by the latent heat of condensation added to it on the way up the windward slopes, and the adiabatic heating on the way down the leeward slope. It is not uncommon for a chinook to develop over the eastern slopes of the Rockies when snow is still on the ground. The snow may literally evaporate before one's eyes: 6 inches have been known to disappear in a matter of a half-hour.

To summarize the adiabatic processes: rising air cools at the rate of about 5½ degrees per thousand feet of elevation gain, so long as it is not saturated. It regains temperature at the same rate when it descends. Once saturated, clouds form and the rate of cooling is reduced to about 3 degrees per thousand feet. If the water is lost by rain or snow, the descending air heats at the dry adiabatic rate. However, the rate of warming will remain at the saturated adiabatic rate so long as liquid water droplets remain in the descending air.

Temperature lapse rates in the atmosphere. The variation of air temperature with altitude is known as the *temperature lapse rate.* Suppose a bubble of air at point A in figure 4-8 is given an upward push, perhaps by the air flowing over rough ground. How will its temperature change as it ascends to a higher level? If it is not saturated, it will cool at the dry adiabatic lapse rate. In figure 4-8, this rate of cooling is represented by the broken line. When the bubble of air reaches level B, it will be warmer than the air around it. (We suppose that only a small bubble is displaced upward; the air layer does not move.) Because it is warmer than the surrounding air at the new level, the bubble will be buoyant: heated air rises. It will continue to rise as long as it is warmer than the air around it at the same level. We can say, therefore, that the air layer in which the bubble is immersed is unstable: vertical

Figure 4-8. Diurnal progression of air temperature profiles near the ground

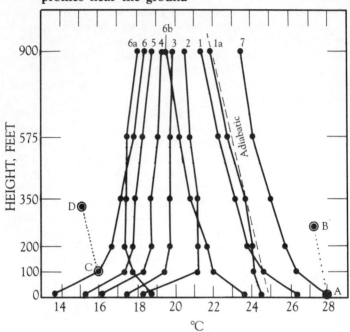

Period	Time
1	1310–1410 first afternoon
1a	2½ hours before period 2
2	1½ hours after sunset
3	Midway between periods 2 and 4
4	Observation closest to midnight
5	Midway between periods 4 and 6
6	Immediately before sunrise
6a	2½ hours after period 6
6b	Midway between periods 6a and 7
7	1310–1410 average the second day

See text for explanation of lines AB and CD. (Adapted from Slade, D. H. (ed); *Meteorology and Atomic Energy—1968.* U. S. Government Printing Office.)

motions are enhanced. The air layer has a lapse rate of temperature greater than the dry adiabatic lapse rate; it is a *superadiabatic* layer.

If the air bubble is in a layer as at C, it will, on being given a vertical shove, find itself colder than the air around it (that is, at D). It will be heavier than the air around it and will tend to sink back down to its original level. The layer at C, incidentally, is referred to as an inversion layer because the temperature increases with height, an inversion of the normal decrease of temperature with height.

Thus, the air in this inversion layer has a stable lapse rate: vertical motions are resisted. The superadiabatic layer, however, tends to enhance vertical motions; it is an unstable layer. An intermediate situation may prevail. If the existing vertical temperature distribution in the air is close to the adiabatic, then vertical motions are neither retarded nor enhanced. The layer is said to be neutral.

An analogy may help clarify these relationships. Suppose we could balance a marble on the top of an inverted hemispherical bowl. Any slight push would send the marble rolling off. But if we invert the bowl and put the marble in the bottom, any push will merely result in the marble returning to its original position. If our bowl were flat and we gave the marble a push, it would neither tend to accelerate away from its original position nor to roll back toward it. These three situations correspond to states of instability, stability, and neutral stability, respectively. Note that we must consider three things in order to determine stability: we need to know the state of the environment (bowl or air layer); we need to give the marble or parcel a push or displacement from its original position; and we need to know the force that develops as a result of the push. In the case of the bubble in the atmosphere, this force is its buoyancy. The same considerations apply if the air layer is saturated; in this case, the neutral lapse rate is 3 degrees Fahrenheit per thousand feet.

Considerations of atmospheric stability are all-important in determining whether such convective phenomena as dust devils, dry thermals, cumulus clouds, and thunderstorms develop. The U.S. National Weather Service maintains a network of upper-air observing stations where twice-daily measurements of the existing temperature lapse rates are made. The observations are made by means of a radiosonde, a balloon-borne radio transmitter that automatically sends temperature and humidity readings, as well as wind speed and direction, at various levels as its ascends through the atmosphere. At a height of 50,000 feet or more, the balloon bursts and the transmitter descends on its own paper parachute. The data radioed back to the ground station are plotted on temperature-height diagrams and a layer-by-layer analysis is made of atmospheric stability.

Weather associated with various stability states. As indicated in figure 4-8, the atmosphere near the ground typically progresses through a series of stability states in the course of a clear day. Many weather phenomena of interest to the outdoor recreationist are intimately associated with this cycle. The powerful heating effect of the sun creates an unstable layer that may extend upward to a height of 3,000 feet or more. In this layer, vertical convection develops readily. Circulations develop, for the air that moves upward must be replaced by downward-moving air. Whatever is in the air is also carried up or down. Smoke from a campfire or a forest fire is carried upward and mixed through the entire layer and thereby diluted. Visibility near the ground improves steadily through the morning hours, reaching a maximum around noon. Isolated mountain peaks are often above the top of the mixed layer, and the climber can readily see the extent of the vertical convection.

If smoke is carried upward and mixed by convection, so also is the momentum of horizontal air currents mixed downward by the countercurrents. As a result, the wind near the surface usually increases during the morning hours as the mixing process brings down faster-moving air. By the same token, the slower-moving air near the surface is transported upward and tends to slow down the air at upper levels. As the day progresses, the wind speed at the surface increases; on the mountaintop, the wind speed decreases (figure 1-4). Because the average speed of the wind increases with height, the wind speed on the mountaintop will usually be higher than the surface wind any time of day.

In figure 1-4, the decrease of wind at low elevations as nighttime approaches is also evident. As the ground cools and stable conditions develop, vertical motions are inhibited and the upper air is decoupled from the surface layers. Accordingly, wind on the mountaintop increases at night as the wind at lower levels decreases.

The vertical convection triggered by unstable lapse rates tends to be rather irregular, with concentrated regions of upward-moving air separated by larger areas of slower-moving, subsiding air. As an upward-moving convection column passes, the wind will pick up and become gusty. As the convection column passes, the wind may die down somewhat, at times becoming nearly calm. When the wind does blow, it is typically gusty and turbulent.

The desert hiker may experience an extreme form of instability: the dust devil or whirlwind, which drains the overheated air from the surface and carries it (and the entrained dust) to heights of several thousand feet. Some may last a few minutes; others may last an hour or more. Their rotation may be either clockwise or counterclockwise, although there appears to be some tendency for counterclockwise motion to prevail, as might be expected from the influence of the earth's rotation. If one ever passes directly over you, the experience will be unforgettable. It will be a long time before you get the sand out of everything.

The rising air in a convection column cools adiabatically. If it rises high enough, the air will cool to the dewpoint, that is, it will become saturated; and a cumulus cloud will form. Typically, these fair-weather cumulus clouds will have flat bottoms and puffy tops. The base of such clouds is at the so-called condensation level and

Figure 4-9. Cumulus

indicates the level at which air from the surface layers reaches saturation (figure 4-9). Moist surface air (that is, air with a high relative humidity) will reach condensation at a lower height than dry air. Thus, the height of the cloud base above the ground indicates how moist the surface air is: the lower the cloud base, the moister the air at the surface.

As the day wears on and the heat input to the ground from the sun decreases, the driving force for the convection diminishes; and the clouds tend to dissipate, leaving the sky clear once again. Once the sun has set, the ground cools more rapidly, the air becomes stable, and an inversion layer develops at the surface. The wind dies down and becomes more regular and smooth; the daytime gustiness disappears. Such is the course of events associated with the typical fair-weather progression of stability and instability at the surface.

Local Winds

During stormy weather, when skies are cloudy and the wind is strong, spatial gradients of the weather elements are diminished. Locations many miles distant will have similar temperatures, humidities, and winds. The weather elements are then under the control of large-scale, intense low-pressure systems; and microclimatic variations are small or nonexistent.

When weak high-pressure systems prevail, sunshine is strong, winds are relatively weak, and microclimates are strongly differentiated. Under these conditions, local wind patterns develop in response to spatial variations in surface heating. In mountainous regions, the topography modifies these winds. Because local winds prevail during periods of quiescent, clear weather, they are the winds we are most likely to find during the best hiking weather.

The basic cause of local winds is the differential heating and cooling of the earth's surface under clear skies and weak pressure gradients. The heated places produce vertical convection currents. Air from nearby cooler spots flows in to replace the heated air, and a circulation develops. The scale of such circulations can vary between the drift that develops from a cool forest to an adjacent clear-cut area to a sea breeze that develops along 1,000 miles of coastline and may extend 50 miles inland. But no matter what the scale, the origin must be sought in local temperature differences.

Slope winds.   Perhaps the most obvious local wind the hiker is likely to notice in the mountains is the wind that blows upslope in the daytime and downslope at night. On a clear night, the earth's surface cools by radiating out more heat than it receives. The air in contact with the cooling surface also becomes cool and thus denser than it was. By contrast, the air between opposing slopes remains warm, for it is not in contact with the cooling surfaces. As a result, the cool air near the surface flows downhill and collects in the valley bottom, forcing the warmer air occupying the space between the valley walls up and outward toward the slopes. A circulation is thus formed (figure 4-10).

Figure 4-10. Air circulation in a valley on a clear night

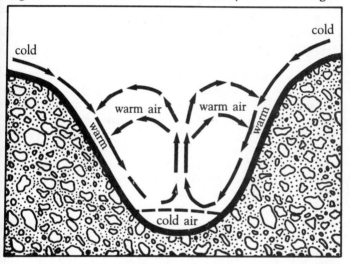

It is tempting to liken these downslope density currents to streams of water flowing downhill. However, there are significant differences. Water will always flow down topographic channels, be they small or large. Air will be guided by topographic channels but may flow across them in a very irregular manner. Downslope winds will flow over obstacles that would divert water currents. So be prepared to find modifications of the general principles that are about to be outlined. Of course, there are always reasons for the variations, and you should always attempt to find explanations in the prevailing meteorological or topographical conditions.

Downslope winds develop most characteristically during

calm nights when the air is clear and dry. Clouds, especially low clouds, keep the surface from cooling rapidly and greatly inhibit the development of the slope wind. Wind speeds are generally rather light, a few miles per hour, and are rather steady, at least on moderate slopes. The wind speed is higher on steeper slopes; but at a certain point, a rather interesting phenomenon, the so-called air avalanche, occurs. Periods of strong, downslope winds alternate with nearly calm periods with a regular periodicity. I have experienced this phenomenon above the timberline in the Sierra Nevada on nights conducive to slope-wind formation. The gusts of cold air developed with a rush, blew steadily for a minute or two, and then subsided to near calm. After about ten minutes, another rush began; the cycle was repeated for several hours. Such rhythmic bursts apparently develop when a pool of cold air develops on a bench or topographic basin. At some point, the pool becomes deep and heavy enough to start flowing down the slope and continues until the pool is effectively drained. Enough time must then elapse until the cold pool is reformed and the cycle starts over again. It appears to be related to the steepness of slope, with steeper slopes having shorter periods. The phenomenon has been observed in the Austrian Alps, with a period of about five minutes; and there is a report of an occurrence in the mountains of central Africa in which the wind reached a speed nearly high enough to blow down a tent. Generally, however, wind speeds are moderate, on the order of 10 mph.

On gentler slopes, the current is more regular, often nearly laminar. Maximum speeds are reached about 60 to 120 feet above the surface, the wind close to the surface being retarded by the friction of the ground. If the slope is forested, the current is much reduced below the canopy but may be blowing at normal speeds above it.

As the air flows toward the valley bottom, it collects there and forms a pool of cold air that increases in depth and decreases in temperature throughout the night, especially if the valley bottom is broad and flat. If, as often happens, the valley bottom is occupied by a meadow with thick mats of vegetation, a true frost pocket may be present. Temperatures may plummet. In one thoroughly studied depression in Austria (the Gstettneralm sink-hole near Lunz), temperatures at the bottom were more than 45 degrees colder than at the top, only 500 feet higher in elevation.

Figure 4-11. Diurnal cycle of air temperatures on a south-facing mountain slope

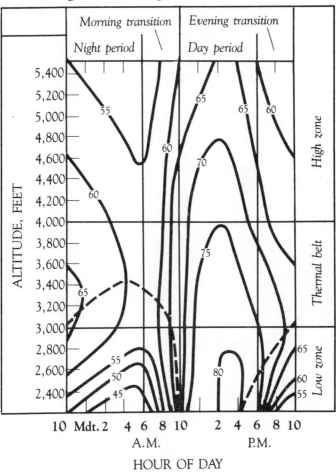

Air temperatures measured 4½ feet above ground in standard weather shelters located at various elevations from the valley bottom at 2300 feet to the ridge top at 5500 feet. Broken line indicates the elevation of maximum thermal belt influence. (Adapted from Hayes, G. Lloyd, *Influence of Altitude and Aspects on Daily Variations in Factors of Forest Fire Danger.* USDA Circular No. 591. 1941.)

If cold air slides to the bottom of slopes and is replaced by warmer air from the free-air region away from the slope, there will be a relatively warm zone along the slope, the so-called thermal belt (figure 4-11). In a typical U-shaped valley, relatively flat on the ridge and in the valley bottom, thermal belts develop regularly along the mid-slopes. On the flat ridge tops, cold air is prevented from flowing toward the valley by the gentleness of the slope. But on the slope itself, the air temperature may remain warmer than above or below, typically about 20 degrees on favorable nights. Gradients tend to be rather sharp, and one may have to move upslope only two or three hundred feet in elevation to find a camping spot that will remain 20 degrees warmer than the valley bottom.

U-shaped valleys with broad bottoms and steep walls develop the most marked contrasts between the valley bottom and the thermal belt. Narrow V-shaped valleys generally have shallow pools of cold air at the bottom and a less well-developed thermal belt. The main reason for this is that the valley walls radiate heat energy to each other, in effect trapping the heat rather than losing it to the vault of cold sky arching over a broad U-shaped valley.

Smaller-scale cold air ponds and thermal belts will develop on benches that sit above the valley bottom, as figure 4-11 indicates. The coldest temperatures will occur at the very bottom of the valley, and the secondary cold-air ponds will not be quite so cold. The relatively flat tops of the ridges are more exposed to the wind which mixes the air and prevents strong inversions from developing, even though much heat is lost to the open night sky.

The transition from the cold air in the valley bottom to the thermal belt is gradual. The center of the thermal belt usually lies about 600-1,000 feet above the valley bottom. The calmer and clearer the night, the higher the thermal belt lies. It also tends to be higher in winter than in summer. Strong winds and a cloudy sky can wipe out the thermal belt completely. At such times, there may be little or no inversion in the valley bottom and temperatures will be uniform along the slope.

During the daytime, as might be expected, the air heated by the slope flows upward. It does not, however, behave like cold air, which hugs the ground; heated air in the daytime tends to peel off and form vertical convection columns. Since there is no place for the heated air to accumulate, no phenomena comparable to frost pockets or thermal belts can develop. Temperature contrasts between valley bottom and valley walls are not great and usually show something like the 5½ degrees per thousand feet adiabatic lapse rate, or even somewhat less. A circulation such as that shown in figure 4-12 will develop.

Because the development of slope winds depends on temperature contrasts, one can expect upslope winds to develop first and strongest on those slopes receiving the greatest amount of solar heating. Thus, east-facing slopes receiving the early morning rays of the sun heat up first and will develop upslope winds at a time when the shaded slope may still be experiencing downslope winds; and the circulation may be more like that shown in figure 4-13. When both slopes become heated by the sun, the more symmetrical circulation of figure 4-12 will develop.

Mountain-valley circulations. As the night progresses, the pool of cold air in the valley bottom builds upward, filling a good portion of the valley. If the valley itself has a significant slope, this mass of cold air starts to slide downward. Again a circulation develops, with cold air moving down the valley and replacement air flowing up the valley at some distance above the surface. The down-valley wind may continue for some time after sunrise, until

Figure 4-12. Air circulation in a valley on clear days

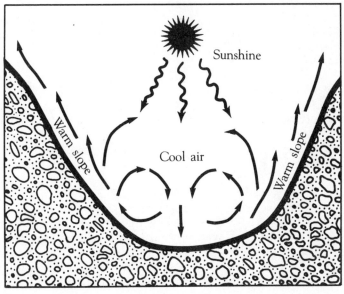

the sun heats the side slopes of the valley enough to initiate the upslope winds discussed in the previous section.

As the valley becomes filled with warm air as a result of the upslope circulation that develops during the day, the entire mass of air within the valley starts to move up. This up-valley movement may occur along the valley slopes as well as at the bottom. Thus, the upslope flow that develops in the morning may turn into an up-valley flow later in the day. It will become more pronounced late in the afternoon, when the sun's rays weaken and the valley slopes start to cool.

During the clear, quiescent weather that is conducive to local wind development, a well-defined circadian rhythm of slope and valley winds can be observed. Figure 4-14 shows an idealized cycle of these winds. Although most features of this circadian rhythm can be observed in many valleys, local topographic irregularities or the force of large-scale weather systems cause many variations. The conditions most favorable to the development of the ideal pattern are long, moderately deep valleys with few twists and turns; presence of a stagnant high-pressure area

Figure 4-13. Air circulation in a valley on a clear day when one slope is shaded

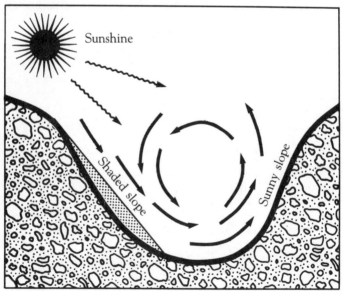

Figure 4-14. Typical cycle of air circulation in a valley

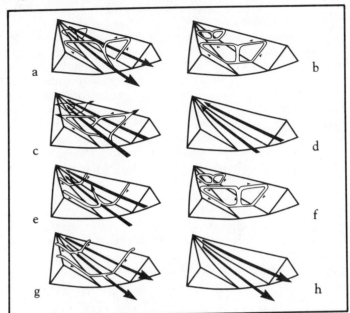

a) Sunrise; onset of upslope winds (white arrows), continuation of mountain wind (black arrows). Valley cold, plains warm.

b) Forenoon (about 0900); strong slope winds, transition from mountain wind to valley wind. Valley temperature same as plains.

c) Noon and early afternoon; diminishing slope winds, fully developed valley wind. Valley warmer than plains.

d) Late afternoon; slope winds have ceased, valley wind continues. Valley continues warmer than plains.

e) Evening; onset of downslope winds, diminishing valley wind. Valley only slightly warmer than plains.

f) Early night; well-developed downslope winds; transition from valley wind to mountain wind. Valley and plains at same temperature.

g) Middle of night; downslope winds continue, mountain wind fully developed. Valley colder than plains.

h) Late night to morning; downslope winds have ceased, mountain wind fills valley. Valley colder than plains.

(Reprinted from Defant, F., "Local Winds" in Malone, T. F., ed., *Compendium of Meteorology*. Boston: American Meteorological Society. 1951, by permission.)

with weak winds; and clear skies both day and night.

A strong gradient wind blowing in the free air above the ridge line may modify the mountain-valley circulation markedly. As one proceeds up the slope toward the ridge line, the effects of the gradient wind become more apparent. Indeed, at some level, the valley wind and the slope wind may be completely overpowered by the upper wind. The strong vertical convection caused by daytime solar heating brings down the faster moving air from above the ridge tops, which then carries along the air near the surface more nearly in the direction and with the speed of the upper wind. As might be expected, this effect is more marked in broad, shallow valleys than in narrow, steep ones.

The interaction between local circulations and the gradient wind is especially well marked in the east-west valleys along the ocean side of the mountains along the Pacific coast. At night, the coupling is weak because of the stable stratification that develops under clear skies. Characteristic downslope and down-valley winds will develop, although at the ridge tops, the gradient wind may be very evident.

During cloudy and windy weather, local circulations will be weak or, more frequently, nonexistent. Nevertheless, the topography itself channels the wind, and there will be strong tendencies for the winds to blow along valley axes. However, the wind may blow up or down the valley without regard to time of day. Deep, narrow valleys will show this channeling to a much greater extent than broad, shallow ones.

Glacier winds.   Glaciers and persistent snowfields create their own local winds. On a sunny summer day, the surface of the glacier will be at the temperature of melting ice, 32°F, and the air in contact with it will be cooled. Surrounding ground and its skin of air will be much warmer. This temperature contrast creates a shallow down-valley wind coming off the glacier. However, as the air flows over the warm ground below the lip of the glacier, it warms rapidly; and so the glacier wind does not usually extend more than one-third mile or so down the valley. It is also a rather thin current of air, a few hundred feet at most, and so does not extend very far up the valley walls.

Glacier winds reach moderate speeds of perhaps 6 to 8 mph and are stronger in the daytime than at night (because the temperature contrasts that drive the wind are greater during the daytime). However, there may be a secondary maximum just before dawn, when the normal down-valley wind is at a maximum. Below the glacier wind, an up-valley wind will blow in the daytime, flowing above the glacier wind as it encounters it. Unlike most local winds, the glacier wind may also be found during cloudy weather as long as the general winds are light. The rule is that whenever temperature contrasts develop between adjacent regions, local winds can arise.

The ecological influence of the glacier wind can be discerned for a considerable distance below the toe of a glacier. Vegetation will be very sparse or nonexistent for a few hundred feet from the lower edge of the glacier; and vegetation will be stunted and deformed for much greater distances. Glacial winds can lower the normal timberline as much as 1,500 feet. The glacier exerts a powerful influence on the vegetation through its modification of the microclimate.

On a much smaller scale, cold air draining from small snowfields or ice caves may also have noticeable ecological effects, delaying the growth and flowering of trees and plants in the cold-air path. The boundaries of these currents are often sharp and the ecological gradients are equally sharp. The same plants only a few feet apart may differ in their stage of development by several weeks.

Sea breezes. The hiker near the shore of any sizable body of water can observe another of the major local wind systems: the daytime sea breeze and its nighttime counterpart, the land breeze. Like all local winds, the sea breeze originates in temperature contrasts that develop between the water and the land. These temperature contrasts develop during clear, quiescent weather because of the sharply differing thermal properties of land and water. Solar radiation that impinges on an opaque solid surface such as bare ground or rock is partially absorbed by that surface. The absorbed energy is turned into heat and raises the surface temperature. This increased surface temperature in turn causes heat to flow into the ground by conduction and into the air by convection. If the ground is covered with vegetation, be it turf or trees, the same processes occur, although modified somewhat. Some of the sun's heat may be used to evaporate water from the vegetation, thus heating the air to a somewhat lesser degree. Less heat flows into the ground because it is effectively shielded from the sun's rays. In either case, a large portion of the sun's energy is used to heat the layer of air near the ground.

On the other hand, the sun's rays penetrate a water surface and are absorbed gradually as they reach lower and lower levels. It is not the surface that is heated so much as it is the entire upper layer of water. The mixing of the surface waters by turbulence helps distribute heat through a thick layer. Evaporation from the water surface also helps keep it cool. Because the heat is spread throughout the absorbing medium, any one part of it is affected much less than the absorbing surface of the solid ground. Sea surface temperatures remain lower than land surface temperatures. It is this surface temperature contrast that creates the sea breeze.

Convection currents occur over the land, but not over the water. Air that rises over land must be replaced: this comes from the nearby water area. A sea breeze develops (figure 4-15).

At night the land surface cools more than the sea surface nearby. The cooled air over the land flows toward the sea. As it flows over the relatively warm water surface, it is heated, becomes unstable, and rises. (This destabilizing effect of the warm water can sometimes be seen when smoke from a campfire drifts in a stable current over the warm water of a lake, where it quickly breaks up into turbulent and irregular motion. This can be seen best just before sunrise, when there is sufficient light to see the smoke drift.) The land breeze is a gentler wind than the sea breeze and is often more noticeable as a drift rather than a breeze. Temperature contrasts between water and land surface are less at

Figure 4-15. Sea breeze circulation

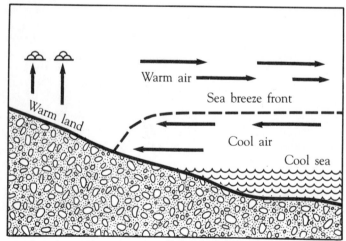

Warm air

Sea breeze front

Warm land

Cool air

Cool sea

night than in the daytime; the resulting wind is thus less at night.

It might be expected from this description that the daytime sea breeze would develop slowly as daytime heating proceeds. It usually does not, however: the sea breeze commonly starts with a rush, bringing lower temperatures and a quick change of wind direction from seaward to landward. It appears to act as a miniature cold front, which, in a way, it is. At night, the cool air from the land collects over the water, gradually building up a pool of relatively cold air some distance from the shore. During the daytime, the land surface becomes heated from the sun more or less uniformly; convective cells develop and a local area of low pressure develops over the land. The cooler air over the water starts to move landward, pushing the warmer air ahead of it. Thus a kind of cool front develops at the leading edge of the landward-moving air.

This so-called sea breeze front travels at a modest speed, perhaps 6-8 miles per hour, and may reach 25 or 30 miles inland before dying out. The sea breeze front thus reaches inland areas much later than it does the shorefront. The farther inland one is, therefore, the later the onset of the sea breeze. Whereas near the shore, the sea breeze may start in mid-morning, 20 miles inland it may start after noon and blow only for an hour or two before the sun's rays weaken and the sea breeze dies out. As might be expected, the air moving inland warms up as its proceeds, and thus reaches inland areas with a considerably higher temperature than it had when it started out at the shore earlier in the day.

The weather conditions that favor the development of the sea breeze are those that make its onset most welcome: hot, sunny days with light prevailing winds, typical air mass weather. A moderate off-land prevailing wind will retard its development. A prevailing wind toward the land will reinforce it. Cloudy skies, of course, will diminish the temperature contrasts between land and water. The sea breeze is not a feature of such weather.

Small lakes can develop miniature versions, although wind speeds will be very low. But large lakes, even those smaller than the Great Lakes, will have well-developed sea breeze circulations.

For the hiker near the coast, then, the sequence is as follows: sometime in mid-morning, depending on distance from the shore, the sea breeze sets in suddenly, dropping the temperature as much as fifteen degrees; the wind blows rather steadily but starts to drop off in mid-afternoon; in the evening, near calm may prevail until a seaward drift starts; the land breeze continues throughout the night but may stop after sunrise, giving way to light and variable winds; in mid-morning, the cycle repeats.

Foehn and chinook winds. A strong south wind off the slopes of the Alps blows across Munich perhaps a dozen times a year, most often in spring and fall. Skies are clear; the air is warm and dry; and the wind blows with a strange constancy. Ideal weather? Perhaps, but the foehn is regarded with apprehension by the Müncheners. Automobile accidents increase; tempers become short; and people are affected by a strange sense of malaise. Indeed, symptoms are so predictable that a medical syndrome, foehn sickness, is recognized. The causes of foehn sickness are still debated vigorously. Is it purely psychological in origin? Or is it caused by some combination of air pressure, humidity, and wind speed?

Whatever the medical results of the foehn may be, the physical causes of the wind are well known. It is a wind that occurs in all mountain regions of the world and is so well defined in its onset and actions that each region has found a special name for it. In North America, the "type" example of the foehn is the chinook of the east slope of the Rockies. On the west coast, the Santa Ana of southern California is a typical foehn wind. Farther north, Mono winds blow down the east-facing slopes of the Sierra Nevada. In the Cascades, the winds are known locally as East winds. The various foehn winds will be described in geographical detail in the individual regional climate chapters. Here we are concerned with the general description of foehn formation.

Stated most simply, foehns form whenever air is forced over a mountain range and down the lee side, so long as the air hugs the surface on its downward journey. The force that pushes a foehn may be a low pressure system, typical with chinooks in the northern Rockies; or it may be a high pressure area, such as drives the Santa Ana of southern California. The process may be a continuous one, in which the air is forced in a stream over the mountains; or it may be discontinuous, in which air forced upward onto a high plateau (as in the intermontane West) remains for several days before starting its downward journey. On its way up the windward side, the air cools dry adiabatically at about 5½ degrees per thousand feet until it reaches saturation. Thereafter, it cools at the moist adiabatic rate, 3 degrees per thousand feet, dropping its moisture in the form of rain or snow. The latent heat of condensation is added to the air on its upward journey. On its way down the lee side, the air warms at the dry adiabatic rate as soon as its passes over the crest, for it is no longer saturated. Thus the air warms rapidly and displays a rapidly decreasing relative humidity, and appears at lower levels as a relatively hot and dry air mass.

Because the cloud bank evaporates quickly as the air slides

down the lee side, the cloud at the crest may appear as a wall, the so-called foehn-wall or crest cloud, to the observer in the lowlands. The foehn wall may not always be present at the leeward edge of a broad mountain range, however.

If the descending air displaces colder air that is present near the surface on the lee side, one may notice a dramatic and rapid increase in the air temperature. Indeed, a rise of 40 degrees in fifteen minutes has been recorded in Alberta. The boundary between cold and warm air may oscillate back and forth, producing the strange sensation of rapid changes in air temperature. If there is snow on the ground, rapid evaporation (actually sublimation) of the snow may occur, as well as melting of the snowpack. The rapid disappearance of the snowpack is so dramatic that locally foehns are often termed "snoweaters."

Foehn winds are generally quite strong, having much higher speeds than the gravity-driven nocturnal downslope and downvalley winds described earlier. Speeds of 25 mph are not uncommon. Because of the high speeds that are typical, foehns may not feel warm, even though they are thermodynamically warm. Because of the high velocities, their cooling effect on humans may be considerably greater than the colder, calm air they displace.

Also, because they are driven by large-scale pressure systems, foehns may blow both day and night. Indeed, nighttime velocities may be even greater than daytime velocities: at night, the downslope gravity wind will reinforce the foehn; in the daytime, upslope winds on heated slopes will counteract it.

The foehn effect is sometimes reinforced by dry subsiding air from high above the earth. This only occurs in high-pressure areas and is common in the Santa Ana of southern California. Surface relative humidities may be extremely low, sometimes less than 10 percent. Such low humidities coupled with high winds produce extremely dangerous forest fire conditions.

Making Use of the Sun

Because of the rather complicated motion of the earth in relation to the sun, the distribution of the sun's energy over the earth varies from day to day. Thus, every place on the earth experiences its own peculiar solar climate (the amount and distribution of solar energy available at that place on the clearest days). The duration of daylight, for example, is a function both of season and latitude. For any given day, that duration is less in southern Canada than it is in the southern United States. Hikers accustomed to long hours of daylight in the summer may be caught short on a winter ski tour when there may be less than half the summertime hours from sunrise to sunset. And, of course, the total amount of energy received from the sun on a winter day may be much less than half that received in summer, even when comparing two bright and cloudless days.

There are many other reasons for the backpacker to be familiar with the solar climatology of a region. For example, choice of a campsite may depend on whether one wishes to wake with the sun and be off early or to sleep past sunrise. I remember vividly trips on the high deserts of the Southwest when the nights were cool enough to warrant a sleeping bag; but once the sun poked above the horizon, it felt as if someone had opened the furnace door. Further sleep was impossible. Lying down on the shady side of some desert shrubs may provide enough shelter from the sun to permit an additional hour or so of comfortable sleep.

The sun is due south at noon (sun time), but the sun can be used as a compass any time it is visible if one knows the correct time. Since we have adopted the convention of referring clock times within a broad longitudinal zone to a "standard" meridian, the sun will be due south at noon only at the standard meridian. All other locations in a standard time zone will show a different clock time at the moment the sun is due south of the observer. The elliptical path of the earth around the sun also throws another small correction into the calculation of true sun time.

Nevertheless, it is possible to prepare a table that lists the times of significant solar events such as sunrise, solar noon and sunset for a particular location. So long as we are not far from that location, our table will show the approximate time of sunrise, solar noon and sunset on our clocks set according to standard time. If we wish greater accuracy, we can correct the tabulated values according to the number of degrees of longitude we are east or west of the location for which the table was prepared. For times of the year that "daylight" time is in effect, an additional one-hour correction is necessary.

For each of the regions described in Part Two, such a table has been constructed for a central location. For most purposes, the tabulated times can be used without correction (except for the daylight time correction). The tables also give the duration of "civil twilight," the period after sunset or before sunrise during which there is enough light to permit outdoor activities by natural light.

Chapter 5

Weather Hazards

Most of the time and in most places, the weather is rather benign. At other times, however, nature treats us to spectacular displays of furious weather; outdoor enthusiasts should be aware of the dangers of such weather events and know how to avoid or minimize them.

Body Heat Balance

The human body must maintain its internal temperature within rather narrow limits for proper functioning. Ten degrees on either side of the normal 98.6-degree temperature leads to irreversible physiological changes and death. Fortunately, the body has numerous mechanisms to maintain a suitable internal temperature. Only when weather stresses overwhelm these mechanisms does the healthy body become dangerously hot or cold.

Hikers are especially vulnerable to these stresses. At home, we can always find another sweater or jacket to put on to keep warm. If the weather is too hot, we can always turn on the fan or drink a glass of cool water to replace body fluids lost by perspiration. In the backcountry, though, we have only what we have carried with us or what we can find around us to protect us from weather stresses. Hypothermia (undercooling) and hyperthermia (overheating) are too often the result of inadequate preparation.

In order to clarify their causes and prevention, I shall review briefly how the body maintains its temperature in the face of a changing environment.

The same mechanisms of heat transfer to and from a surface discussed in the previous chapter operate to transfer heat to and from the human body. In determining the heat budget of the body, however, there is one additional source of heat: the internal oxidation of food reserves, which produces metabolic heat. It is this internally produced heat that enables the body to maintain a temperature higher than ambient.

Thus, the body is a kind of heat engine that produces work. But this work also involves the generation of heat: the greater the work output, the greater the amount of heat produced. Because the body must maintain a nearly constant internal temperature, this heat must be eliminated. The primary mechanism for transferring heat from the body's core to the outside is the maintenance of a temperature gradient between the core and the skin. Normal skin temperature is several degrees lower than internal temperature, and this results in a steady flow of heat outward.

The heat budget of the body can be described as follows. Heat is transferred to or from the skin by convection, depending on whether it is colder or warmer than the air next to it. The faster the air blows across the skin, the greater the rate of heat transfer. (See the discussion of windchill, below.) The skin loses heat by evaporation of water from its surface. Even when the skin appears dry, there is a certain amount of moisture loss from it. The body also loses heat by evaporation of water from the lungs and nasal passages.

Heat is also gained or lost through radiative exchanges. Solar radiation and thermal radiation from sky and ground are absorbed by the body. This process tends to heat it up; but the body also emits thermal radiation by virtue of its temperature, and this helps cool it. Heat may also be conducted to or from the body through points of contact with the ground. This is relatively unimportant when only shoes contact the ground but may be very noticeable when lying on bare ground in an easily compressed down sleeping bag.

Heat is produced internally in substantial amounts by the oxidation of food reserves. Even a resting person produces heat at the rate of 100 watts. An active hiker may produce more than six times this amount. If this heat were not dissipated, the core temperature would rise to lethal levels. The heat is transferred to the skin in two ways: by conduction through body tissues and by the transport of heat in the bloodstream, which is constantly circulating between the core and the skin. The control of blood flow is thus a major control over body heat loss. For a given skin temperature, a large flow of blood to the skin means a large heat loss. If the core is losing heat at such a rate that its temperature starts to fall, the blood vessels near the skin become constricted (vasoconstriction), and the amount of heat carried to the skin by the blood is reduced. Although this increases the temperature gradient from core to skin and thus increases the amount of heat conducted to the skin, the net result is a reduction of heat loss from the core. The body has a marvelous set of mechanisms to regulate its heat balance and maintain the core near its optimum temperature of 98.6°F. Under any particular set of environmental conditions and level of internal heat production (metabolism), the body adjusts the heat balance of the skin in an attempt to maintain a constant internal temperature. If it is too hot, the skin oozes sweat to promote cooling by evaporation. If it is too cold, it reduces blood circulation to the skin to reduce heat loss from the core. Goose pimples are a mechanism to roughen the skin's surface and reduce the airflow over it and consequently reduce heat loss. We can help or hinder the process by the clothes we wear. Living in a cold, temperate climate as we do, we must help the body maintain its temperature through appropriate skin coverings.

Windchill. When the air is colder than the skin, heat is lost from the skin by convection. The rate of heat loss is proportional to the rate of airflow over the skin. That is, the harder the wind blows, the greater the rate of heat loss and the colder the skin will be. Because our sensation of temperature depends in large degree on skin temperature, we expect to perceive the same degree of comfort at various combinations of wind and air temperature that produce the same skin temperature. More important, we expect these combinations of wind and air temperature to produce the same heat loss from the body.

The so-called windchill index or (preferably) windchill temperature was developed by Paul Siple during Antarctic expeditions. It is defined as the temperature that essentially still air (about 3 mph or less) would have to have in order to produce the same heat loss from skin exposed to a given combination of wind and temperature (table 5-1). For example, at an air temperature

Table 5-1. **Windchill temperatures**

| Air temperature °F | Wind speed, mph | | | | | | |
| | 3 | 5 | 10 | 15 | 20 | 25 | 30 |
			Windchill temperature, °F				
60	60	56	50	48	46	44	43
55	55	51	45	41	38	36	35
50	50	45	37	33	30	28	26
45	45	39	32	27	23	19	17
40	40	34	23	20	15	11	9
35	35	29	18	12	7	3	0
30	30	24	12	3	−1	−5	−9
25	25	17	3	−4	−9	−15	−17
20	20	12	−4	−13	−18	−23	−25
15	15	5	−11	−20	−24	−33	−35
10	10	0	−17	−27	−33	−38	−41
5	5	−6	−23	−33	−39	−46	−50
0	0	−13	−29	−40	−46	−53	−57
−5	−5	−17	−37	−47	−53	−61	−66
−10	−10	−23	−42	−53	−60	−68	−72
−15	−15	−28	−48	−60	−67	−77	−81
−20	−20	−33	−55	−67	−76	−85	−89
−25	−25	−40	−62	−75	−83	−93	−98
−30	−30	−45	−67	−81	−92	−100	−105

of 20°F and a wind speed of 15 mph, the windchill temperature would be −13°F. That is, heat loss from the face or a bare arm exposed to these conditions would be approximately equal to that from an exposure to a 3 mph wind when the air temperature was −13°F.

Of course, we do not go around unclothed when the wind is blowing a gale and the thermometer is hovering near zero. Nevertheless, the windchill temperature is a good guide to the amount and kind of clothing that must be worn to provide adequate protection.

Hypothermia. The rate of heat loss from the core to the skin depends on the temperature difference between the two. Because the core is maintained at a constant temperature, it would appear that the lower the skin temperature, the greater heat loss. This is true of the heat conducted to the surface, but heat loss, as we saw above, can be reduced by vasoconstriction. If the core temperature continues to decrease, the body's metabolism will speed up, producing heat. Exercise also increases the metabolic rate, but much of this heat is produced in the leg and arm muscles and does not help much in keeping the core warm. Indeed, as the cold environment drains more heat from the body, circulation to the extremities is almost completely cut off. The temperature and heat content of the arms and legs are sacrificed, as it were, in order to maintain a suitable environment for the vital organs. If the body cannot produce enough heat to maintain its normal core temperature, the body starts to cool down. If the cooling is not checked, death eventually results.

This sequence of events, known as hypothermia, does not require very low ambient temperatures for its initiation. All it requires is that the heat loss from the body be greater than that produced by it over a period of several hours. Indeed, hypothermia is probably more common with ambient temperatures in the fifties, especially if combined with rain and wind, than it is at temperatures below freezing. At the lower temperatures, we are more likely to be dressed appropriately and be more aware of the dangers.

The first symptom of incipient hypothermia is shivering. This is one of the methods the body has to produce heat: the muscular contractions convert stored energy into frictional heat. But this does not help too much because the heat is produced mostly in the extremities, whereas it is needed most desperately in the body core. At this time, the core temperature is only 2 or 3 degrees below normal. As the core temperature falls to 92° or 91°, shivering becomes more violent and may occur in spurts. Muscular coordination is affected; speech may be difficult and somewhat incoherent. At a core temperature 5 degrees lower, shivering ceases and muscles become stiff. Muscular coordination is severely affected; thinking and judgment are impaired. At a core temperature below 80°, the victim of hypothermia becomes unconscious and heartbeat and circulation are slowed. With continued cooling, the victim soon dies.

Hypothermia can be prevented simply by reducing the heat loss from the body when that loss is greater than the rate of internal production. But hikers often are not aware of incipient hypothermia until they start shivering violently. Too often, a hiker will start a climb in the valley clad in tee shirt and shorts. This may be appropriate in the low-elevation forest. But above the timberline, with the hiker enveloped in a cold, drizzly fog, such clothing may be completely inadequate to stem the increased heat loss. The remedy is to be constantly aware of changes in heat loss conditions—increased wind, rain, and lower temperatures—and to adjust the amount of clothing worn before symptoms appear.

Heat balance considerations also dictate the treatment for hypothermia. Because the body is cold already, putting the victim alone in a sleeping bag does little good. Heat must be transferred to the victim as rapidly as possible without causing damage to the skin. Heat can be effectively transferred to the victim from a warm person (or persons) by direct skin contact, all inside a sleeping bag or blankets to conserve heat. Hot drinks laced with sugar help if the victim can swallow. Radiated heat from one or more fires may be appropriate, but care must be taken not to burn the skin from too much radiation. The rewarming process may take several hours and must be continued until all symptoms have disappeared and the body temperature is back to normal.

Hyperthermia.   A resting individual without clothing is comfortable when the ambient air temperature is about 84°F as long as there is no net gain or loss of heat by radiation and the air is still. At this temperature, heat loss from the skin just balances the heat produced by metabolism, and the body has to invoke no mechanism for increasing or decreasing either internal heat production or external heat exchange. This is the condition of thermal neutrality. If the ambient air is warmer than this, the body can no longer get rid of the heat; and the body undergoes certain physiological changes.

Perhaps the most obvious change is that the sweat glands are activated and the perspired moisture, upon evaporating from the skin, cools it. Thus, the heat balance is redressed. An increase in wind speed across the skin will not only increase evaporation but will also convect more heat from the skin. Blood vessels may dilate, increasing the circulation of blood and thus carrying more heat from core to skin. By contrast, exercise will increase the amount of heat produced internally, placing an additional burden on the heat transfer mechanisms.

With extremely high air temperatures, those above the normal skin temperature of about 90°F, a paradox occurs. Although wind promotes the evaporation of sweat and thus tends to cool the skin, the air, being warmer than the skin, actually tends to warm the skin. Because there is a limit to the rate at which evaporation of sweat can occur, the faster the wind blows, the hotter the skin becomes. Under these conditions, evaporation of sweat is the only way excess body heat can be dissipated. Sweating depends on an adequate supply of moisture. An active backpacker can sweat as much as several quarts of water in a day, so this much must be replaced. One should drink lots of water; it is better to drink small amounts frequently than to gulp huge quantities all at once. But the old wive's tale about not drinking in the desert is a fallacy that can lead to serious consequences. The best way to avoid hyperthermia is to drink plenty of water and do those things that will minimize heat gain to the body and promote evaporation of sweat: wear a broad-brimmed hat; rest in the shade; wear loose clothing.

Bioclimatic Index

There are four major meteorological controls over the heat balance of the human body: air temperature, humidity (insofar as it controls evaporation), wind speed, and solar radiation. It would be useful if these factors could be integrated in some way that would indicate the degree of comfort or discomfort a hiker might experience. In order to be applied to mountain regions where hikers are likely to go, it would have to be based on meteorological data available for such areas. Unfortunately, for most backcountry areas, only temperature and precipitation data are available. Wind measurements are rare and sunshine data are nonexistent.

A bioclimatic diagram that utilizes temperature and precipitation (as a surrogate for humidity) was developed by the U.S. Army during World War II. An adaptation of this diagram is shown in figure 5-1 and is used in the regional climatologies in the second part of this book. In figure 5-1, monthly mean temperature is plotted against monthly precipitation, one point for each month of the year. The twelve points are connected to form a climogram that indicates visually the progress of the bioclimate throughout the year.

Each cell in the figure is labeled with a subjective classification of the bioclimate. Thus, a month with an average temperature of 75°F and a rainfall of 2 inches would be perceived as hot and dry, whereas a month with the same average temperature but a rainfall of 10 inches would be perceived as hot and wet. The lines above 32°F slope downward from left to right in recognition of the fact that high humidities (as indicated by large rainfall amounts) reduce evaporative cooling from the skin to such an

Figure 5-1. Bioclimatic diagram

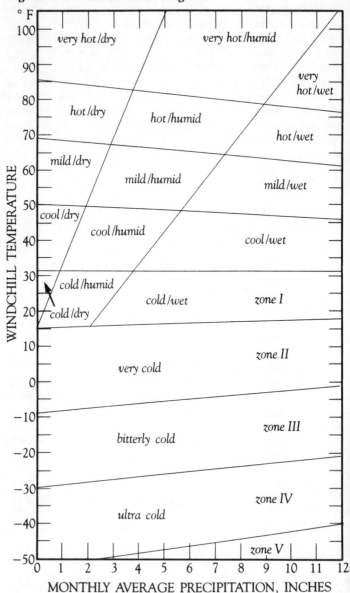

extent that the same temperature feels hotter in a humid region than in a dry one.

Below the freezing level, humidity has a slight opposite effect and so the lines slope upward to the right. That is, the same temperature feels colder if the air is more humid. At very low temperatures, there is little humidity effect and the comfort zones are labeled without a humidity component.

At temperatures below about 60°F, windchill can be important. Accordingly, it is appropriate to calculate windchill temperatures if wind data are available and use this in place of air temperature. Note that windchill is primarily a factor in exposed places. In the shelter of a dense forest, winds rarely exceed a few miles per hour; and it is appropriate to use actual air temperature.

Figure 5-1 can be used to select clothing that should be worn and the precautions that should be taken in inclement weather, especially at low windchill temperatures. The zones and their interpretation, based on a study by the Atmospheric Environment Service of Canada, are as follows:

zone I: cold but comfortable with normal outdoor clothing and precautions

zone II: very cold; work and travel become uncomfortable without proper clothing (that is, all skin areas protected from direct wind, with adequate insulation to prevent excessive heat loss)

zone III: bitterly cold; work and travel become hazardous; heavily insulated outer clothing necessary

zone IV: ultra cold; exposed flesh will develop frostbite after moderately long exposure; heavy outer clothing mandatory; face protection desirable

zone V: unprotected skin can freeze in one minute; multiple layers of clothing mandatory; adequate face protection mandatory; work and travel alone not advisable

The diagrams in the regional climatologies are based on average conditions; the conditions prevailing at any particular time may be better or worse. For example, the summit of Mount Washington in May has an average windchill temperature that places it in zone II, very cold. But it is not uncommon for May windchill temperatures to reach the dangerously cold levels of zone IV or even zone V. Thus, the climograms, especially for exposed places and above the timberline, should be interpreted conservatively. Bright sunshine can add a considerable amount of heat to the body, enough, perhaps, to raise the level of the climogram one band: very cold becomes cold; hot becomes very hot.

The guiding principle for generalizations about the most comfortable appropriate clothing for various kinds of bioclimate is that the clothing should mediate heat exchange to and from the skin so as to maintain the skin at a comfortable temperature. In a *very hot, dry climate*, the problem is to prevent heat gain to the skin by convection and radiation while permitting evaporation of sweat. A relatively thick though loose, light-colored garment fits these specifications. The light color reflects the sun; the thickness provides insulation to minimize heat gain from the hot air; and the loose fit permits evaporated sweat to find its way out. The Bedouin's desert robe is no accident. The head should be covered, preferably with a broad-brimmed hat.

In a *hot, moist climate*, the goal is to prevent as much solar radiation from reaching the skin as possible while permitting a free flow of air. A thin, open-weave white shirt is appropriate here, along with thin, light-colored pants or skirt. Underclothing should be kept to a minimum.

For *cold, dry climates* (below freezing), the minimum amount of clothing consistent with comfort and maintenance of a dry skin is essential. This implies layers of clothing that can be readily donned or shed. A light, nearly windproof outer garment reduces wind penetration; but it should be loose enough to permit circulation of moist air to the outside. Sweat is the great enemy of body heat conservation in cold weather. As evaporated moisture works its way through a thick garment, it eventually reaches a low enough temperature to condense into liquid water. Then it wicks its way back to the skin, where it again requires heat to evaporate it. Thus, the skin should be kept cool enough so it does not become wet with perspiration. This prevention of wicking is the principle behind the plastic waterproof socks next to the skin. Water vapor can then no longer leave the skin to condense and be wicked back. This really works, although it keeps the feet damp, even wet. A warm head covering and perhaps a face mask are also indicated. Thick mittens of wool or down plus overmitts are necessary.

For *cold, wet conditions*, wear clothing similar to that for cold, dry conditions, with the substitution of a waterproof outer garment. This should nevertheless be loose enough to permit circulation of air from the clothing underneath to the outside. Wool is especially appropriate for both underwear and outer garments because it maintains much of its insulating quality when wet. A knitted wool hat and wool gloves or mittens are necessary.

Lightning Storms

Lightning is certainly the most awesome hazard faced by outdoor recreationists. Fortunately, the chances of being struck by lightning are rather low even in the most obvious danger spots, such as mountain peaks, open ridges, and water bodies. Consider how few trees in such areas show evidence of lightning damage. In the United States, annual deaths from lightning number about three hundred, about the same as deaths from snake bite.

The thunderstorm is the product of a series of events that triggers vertical motion in the atmosphere. These vertical motions may have upward velocities of more than 60 mph in a well-developed storm. By contrast, the vertical velocities associated with the large-scale sliding of one air mass over another, as described in chapter 3, may be less than 1 mph.

The air-mass thunderstorm begins with the development of a small, puffy cumulus cloud, which forms when air heated by contact with sun-heated ground becomes buoyant enough to rise like a balloon. As long as the air is dry, it remains invisible. Over desert areas, it may become organized into dust devils and is made visible by the entrained dust. More typically, the rising air cools by expansion until it reaches saturation, at which point the moisture starts to condense into cloud droplets. The air continues to bubble upward, forming the visible domed top of a cumulus cloud. The bases of these clouds are flat and nearly all at the same level, for the air rising from the surface of the earth has about the same moisture content in a single air mass. In a relatively dry continental polar air mass, *cumulus humilis* clouds (figure 4-9) continually evaporate and reform; few will show substantial vertical development. In moister air masses, however, and especially if there are unstable layers aloft, the upward motion may continue for many thousands of feet, forming the towering clouds known as *cumulus congestus*.

These clouds are composed of liquid water droplets, as their sharp outline indicates. Upward velocities are generally low, although in the upper portions they may reach 10 or 15 mph. The effect on ground conditions beneath the cloud is slight. Indrafts must be present in order to replace the air moving upward, but the speeds are low and may be obscured by other air circulations, such as valley breezes. The ground is partially shaded and cooled, and this tends to reduce the likelihood of additional convection.

It may take only a half hour for a cumulus cloud to develop into one 10,000 feet high; an average vertical velocity of only 4 mph would be required. In this time, coalescence of cloud droplets into larger rain drops can occur, with some light precipitation developing.

If the cloud top has reached into the freezing level, profound changes start to occur in cloud structure. The updrafts become organized into a turbulent chimney in which velocities may reach 30 or 40 mph. Cloud droplets start to freeze into ice crystals, a process that triggers the formation of significant precipitation. The level at which this occurs can often be seen from the ground by observing the changeover from the hard outline of a water cloud to the fuzzy outline of an ice-crystal cloud (figure 5-2). As precipitation develops and falls through the cloud, it creates a downdraft. This results partly from the downward momentum of the droplets themselves and partly from the cooling of the air by partial evaporation of the droplets. The downdraft accompanies the rain and occurs in the forward portion of the storm. Below the cloud, it spreads out laterally. Ahead of the rain area, it produces the rush of cold air that often precedes the onset of rain by a few minutes. Eventually, the supply of moisture for the cell is used up, the updraft weakens, rainfall diminishes, and the downdraft dissipates.

A mature thunderstorm may be composed of several such cells, each going through the cycle at different times. Thus, the

Figure 5-2.   Cumulus congestus

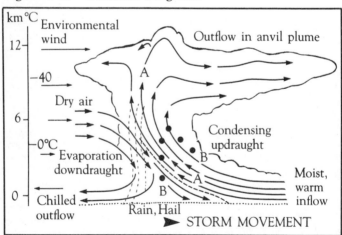

life of a single cell may be about one hour, but the thunderstorm may last for several hours.

The upward-moving water droplets and ice crystals carry with them electric charges, creating a region of positive charges in the upper portions of the cloud and leaving the bottom portion negatively charged. The negative charge near the cloud base induces a positive charge in the ground. When the potential difference builds up sufficiently to overcome the resistance of the air, a rapid discharge occurs: lightning strikes.

The actual discharge between cloud and ground takes place in two stages. First, a "leader" works its way downward to the ground, ionizing the air in its path and reducing the effective path resistance. Then one or more return strokes flash upward to the cloud. The average number of return strokes in a single discharge is four. As a result of the discharge, the negative ground becomes positive and ground currents develop to redress the balance. These ground currents are potentially more hazardous to the hiker than are the air discharges, for they cover a larger area. Other discharges occur between oppositely charged parts of the cloud, producing so-called cloud-to-cloud lightning. These within-cloud discharges outnumber cloud-to-ground strokes by two or three to one.

The noise generated by a lightning stroke comes from the explosive heating of the air channel coursed by the discharge. This creates a compression wave that radiates outward from the path at the speed of sound, about 1 mile in five seconds. The rumble of distant thunder is caused by multiple refraction from the ground surface and upper inversions and partly from the orientation of the path with respect to the hearer. Different portions of the stroke path will be at different distances from the hearer, so the sound will arrive at different times.

The sliding of moist air over a warm front occasionally produces thunderstorms, too, but these are generally the least severe of all types. Air forced over the steep leading edge of a rapidly advancing cold front often produces squall lines, many of which have severe lightning and are occasionally the source of nature's most violent storm, the tornado. Fortunately for the wilderness hiker, these are less common in the mountain regions than they are in the Great Plains.

Air mass thunderstorms occur most often in mid-afternoon, as might be expected from their origin in daytime solar heating of the ground. In the Great Plains, however, their occurrence peaks at night (figure 5-3). In Florida and northern New Mexico, thunderstorms occur on as many as seventy to eighty days per year. Many thunderstorms in the central Rocky Mountains are accompanied by hail.

There are a number of precautions hikers and climbers caught in a lightning storm can take:

1. Avoid or leave exposed ridges and peaks. Even a few yards off the ridge is better than the ridge itself.
2. The best stance is a crouch, with feet close together. This minimizes the opportunity for ground currents to find a path through the body. Crouch on a sleeping pad, if available; but keep it dry. Use the same stance in an exposed tent.
3. Avoid single large trees. The safest place is an opening in the trees or a clump of smaller trees in a dense forest.
4. Do not huddle or stay close together. Scatter so if one person is injured, the others can help.
5. Stay out of shallow caves or overhangs. Ground currents may jump across the openings. A deep, dry cave offers more protection; but do not lean against the walls. Adopt the feet-together crouch.
6. Avoid a depression with a stream in it.
7. In a shelter that does not have lightning protection, lie on a wooden bunk inside a sleeping bag. Avoid metal bunks.
8. If in a boat, go below or crouch in the middle of the boat.

There is much more danger from ground currents than from a direct strike. Even a direct strike is more likely to stun than kill outright. Permanent injury is rare, but it is important that someone be available to administer first aid. Artificial respiration is frequently necessary. Therefore, it is extremely important that members of a party stay at least 30 feet apart.

Hurricanes

Fortunately for hikers, hurricanes are rare in recreational areas. Most hurricanes affecting the North American continent are spawned in the Atlantic or Gulf of Mexico and are carried inland on the currents of the Bermuda High. After moving inland, their intensity usually diminishes and the major effect is a period of heavy rain and high but not destructive winds.

A study by the U.S. National Weather Service indicated that in the Appalachian Mountains, a tropical storm caused destruc-

Figure 5-3. Hour of maximum frequency of thunderstorms, local solar time

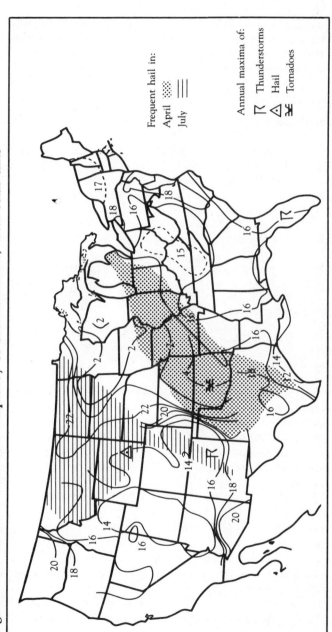

Frequent hail in:
April :::::
July ≡≡≡

Annual maxima of:
ᛘ Thunderstorms
△ Hail
ӂ Tornadoes

tion on the average of less than one year in ten. Immediately along the coast, the storm frequency was somewhat greater—about one year in five. At Cape Hatteras in North Carolina, the frequency was one year in three. Few hurricanes affect the west coast—none at all north of Los Angeles. A few years ago, however, a hurricane did strike the southern coast of California. Its heavy precipitation area moved into the southern Sierra Nevada, causing a rare heavy September snowfall at high elevations.

It is difficult for the isolated hiker to foretell the approach of a hurricane. The cloud shield is similar to that of an approaching warm front. The best indication is a steady and dramatic fall in atmospheric pressure, which can only be monitored if one has a pocket altimeter or barometer.

Snow Avalanches

Few natural hazards that confront the skier are more frightening than snow avalanches. Yet it is a hazard that few skiers think about, much less plan for. In developed ski areas, there is a comforting apparent safety in numbers. Surely where one has skied, another can follow with safety. Even in the backcountry, most open slopes appear secure. But when snow, slope, and weather conditions are right, slopes can avalanche without warning, cascading millions of tons of snow in a great churning mass onto the slopes below.

Every snow-country recreationist, whether skier or snowshoer, downhill skier or backcountry tourer, should know enough about snow and avalanches to avoid danger. In a ski area, it may be sufficient to treat a "Closed—Avalanche Hazard" sign as if it were a 10-foot fence. In the backcountry, the skier must know how to evaluate snow conditions and avalanche weather sufficiently well to minimize the risks. The risks cannot be completely eliminated, however; every backcountry skier in avalanche-prone areas, no matter how cautious and experienced he or she may be, must accept the risks and act accordingly.

Newly fallen snow is one of the most unstable natural substances on earth. When snow falls on a slope, part of the force of gravity acts to press it against the slope and hold it there by friction. The other part of the gravitational force acts parallel to the slope and tends to pull the snow downhill. Simply put, avalanches occur when the downhill forces overcome the forces that tend to keep the snow in place.

Avalanches are of two general types: loose-snow avalanches and slab avalanches. Their causes and behavior differ.

Loose-snow avalanches. On steep slopes, those greater than 45 degrees, new snow generally does not have enough initial cohesiveness to remain on the slope to any depth; and avalanches occur frequently in small masses called *sluffs*. Most of these are small and do not pose great danger to skiers or mountaineers, although they can upset skiers or carry them into a dangerous situation. Only occasionally are they large enough to present a serious hazard. If the fresh snow is wet, even slopes as gentle as 15 degrees have been known to avalanche. Such snow has very little cohesiveness; the free water in it acts to lubricate the mass. Wet-snow avalanches can be dangerous. In the Pacific coast mountains, where new snow is often wet and heavy, such snowslides have destroyed life and property.

The process of sluffing tends to stabilize slopes, removing accumulations from steep slopes and redepositing the snow in gentler terrain. However, sluffs may trigger avalanches on slopes that normally are not steep enough to slide. Slopes that are shedding small sluffs during a storm should be treated with great caution. Because sluffing will often occur with regularity during a storm, skiers can time their crossing to avoid them.

After a storm, sluffing ceases but larger loose-snow avalanches may occur, triggered by natural causes or by the weight of a skier (or by the artillery fire of snow rangers). They start from a point source: a small mass breaks loose and moves downhill, triggering the further sliding of the snow below it. The path propagates sideways as well as downhill and usually takes the form of an elongated teardrop.

Slab avalanches. Snow structure changes continually from the moment it is deposited on the surface. Either through the action of internal changes in the crystalline structure of the snow or through packing by the wind, snow typically develops a cohesive, slablike structure. When the downhill forces become greater than those holding the snow to the slope, a slab breaks loose, triggering a so-called slab avalanche.

Slab avalanches, although less frequent than loose-snow avalanches, are far more dangerous. As the name implies, they start with a large slab breaking loose from the slope and sliding downhill. The initial slab may be as small as 1,000 square feet in area or as large as 100,000 square feet and from 1 to 3 feet thick. The slab quickly breaks up into smaller blocks; their size depends on the cohesiveness of the snow and the roughness of the terrain over which the snow is sliding. The force of the avalanching snow triggers more fractures, and the slide is propagated downward, releasing the pent-up potential energy contained in the metastable snowpack.

A slab avalanche leaves a fracture wall at its upper end. This wall may extend across a slope for many feet; some have been measured to extend for half a mile. The sides of the snow slab will also fracture along the fall line, producing a more or less rectangular pattern on extensive slopes. In gullied terrain, topography may channel the snow into the gullies, forming a track hundreds or thousands of feet long.

The starting zone of slab avalanches must be steep, but not too steep: between 30 degrees and 45 degrees. At greater inclinations, slopes are generally kept free of snow by sluffing; at lower angles, the downhill forces are not great enough to initiate sliding even though fracturing may occur. In addition to steepness of slope, the starting zone must receive large quantities of snow. This often occurs on the lee side of ridges, especially in bowls and gullies. The most active starting zones are gullies bounded at the top by horseshoe ridges or cliffs.

Once started, a slab avalanche will continue its progress downhill as long as the slope has an inclination greater than about 15 degrees. At the bottom of a slope, the avalanche usually spreads out and grinds to a halt in a runout zone. Occasionally, the snow will be moving with such momentum that it runs across a valley and part way up the opposite side.

Metamorphosis of the snowpack. From a fluffy, unconsolidated mass of new fallen snow particles, the snowpack gradually metamorphoses into a denser, more cohesive structure. The structure may be modified further as the temperature changes and as rain or more snow is added. With these changes come changes in the likelihood the snow will avalanche, and changes in the type of avalanche that may occur. Knowledge of the more important of these changes will help the skier in evaluating avalanche hazard.

Individual crystals of snow pressed together in an accumulating snowpack quickly lose their crystalline structure and become grainy. Grains of snow that are in contact become joined by bridges of ice, resulting in a knitting together or "sintering" of the snow mass. This process continues as long as the temperature of the snowpack remains more or less constant and uniform, but most of the increase in cohesiveness occurs within the first day or so after a storm. As the snowpack strengthens, the likelihood of loose-snow avalanches diminishes.

If the surface of the snowpack becomes much colder than the deeper layers, the temperature gradient that develops within the snowpack creates a layer of hoarfrost, so-called "depth hoar." This layer contains large, coarse grains that have relatively few points of contact. It thus has much less cohesiveness than the finer-grained snow from which it is formed. It normally takes about two weeks for this layer to form. The result is a layering of the snow with two layers of fine-grained cohesive snow separated by a lens of weak depth hoar. When this layer becomes too weak to hold the overlying pack in place, and if the downhill force is great enough to start a fracture (perhaps triggered by a skier traversing the slope), a slab breaks loose and the avalanche is on its way.

Depth hoar is most likely to form in early winter when the snowpack is thin and unconsolidated and the ground beneath is comparatively warm. An early snowstorm followed by several weeks of clear, cold weather can set the stage. These conditions occur most often in the interior mountains. In the Cascades and the Sierra Nevada, frequent snowstorms and infrequent interstorm cold periods generally prevent the formation of depth hoar.

There is another mechanism that creates a weak layer in the pack that can lead to avalanches: a melt-freeze cycle. When meltwater or rain penetrates a snowpack, large grains are formed at the expense of smaller grains. As the temperature drops below freezing, the large grains become cemented together to form yet larger units. After several melt-freeze cycles, the well-known granular corn snow develops. In the melt phase, the strength of such snow is extremely low, as skiers who have sunk to their knees in spring mush can attest. When it freezes, however, the mass can become an almost solid block of ice. If a solid crust forms below the surface, meltwater can puddle on it, acting as a lubricant. The upper snowpack is ready to slide.

Avalanche weather. Although the relationship between weather conditions and the occurrence of avalanches is far from perfectly understood, there are some conditions known to favor avalanche development. Snowstorms producing large accumulations of cold, fluffy snow will produce almost continual sluffing on steep (30-degree to 45-degree) slopes. If the storm is accompanied by high winds (generally greater than 15 mph), snow may accumulate to great depths on the lee side of exposed ridges. The wind may further pack the snow, forming a cohesive slab atop a layer of unconsolidated snow. This combination creates a dangerous situation: the slab may let go of its own weight. Heavy snow laid down at temperatures near freezing with little or no wind tends to be more stable.

A change to rain during a storm may produce an unstable condition: a heavy, well-lubricated wet layer on top of dry, unconsolidated snow. Falling temperatures during a storm indicate a trend toward snowpack stability.

Once a storm is over and storm-related avalanching has ceased (usually within twenty-four hours), weather also greatly influences snowpack stability. Clear, cold weather leads to the formation of depth hoar, creating slab-avalanche conditions. This condition is most likely to occur in early and midwinter in the interior mountains, less likely in spring and in the coastal ranges.

Heavy rain on an old snowpack may lead to instability by adding weight and by lubricating the upper layers. A particularly dangerous situation arises if there is a surface within or beneath the pack over which a slab avalanche can slide. Because such hidden surfaces can only be determined by digging snow pits and studying the snowpack stratigraphy, all rain-soaked snow must be treated with caution.

Spring thaws after a heavy, late season snowfall can have similar effects. The cycle of cold nights and sunny, warm days produces the corn snow much prized by spring skiers. But corn snow is well lubricated and will avalanche easily, even on moderate slopes. Such conditions are most dangerous in the afternoon, when the sun's rays are readily absorbed by a granular surface.

Avalanche precautions. Safe winter travel in avalanche-prone backcountry depends on the traveler's ability to evaluate terrain, weather, and snowpack conditions. There are several steps one can take to avoid avalanches, though some degree of hazard is inescapable. Parties must be prepared to revise planned routes and even to abandon tours if unacceptable hazards develop.

When planning a winter ski tour, obtain information on avalanche-prone areas from local offices of the appropriate land-management agency, typically the U.S. Forest Service or the National Park Service. Just prior to departure, check the weather forecast. In many parts of the country, avalanche warnings are prepared by the forest service and the national weather service and disseminated by the public media. At ski areas, the ski patrol office is a good source of information on local conditions and closures. And, of course, study the sky and evaluate the probable weather in the next day or two.

Always be on the alert for hazardous conditions in avalanche country, even in low, timbered valleys. Avalanches starting at high elevations may course through timberland that has remained in place for dozens, sometimes hundreds of years. Constantly be on the lookout for obvious avalanche paths and avoid them like the plague. Clear swaths running from the timberline well down into the timber are indications of frequent, even annual slides. Older slides may not be so obvious, however, with a young stand of timber recapturing an old path. These may often be identified when viewed from a distance, perhaps from a nearby ridge. Gullies often collect avalanches and should be avoided. On open slopes above timberline, avalanche paths are obscure, although they may sometimes be discerned by observing the paths entering the timber below.

Ridges usually provide a safe route, but take extreme care to avoid cornices. Stay on the windward side of the crest, away from the apparent ridge crest, which may be nothing more than the top of a snow overhang only a few feet thick. With poor visibility, such ridges can be very hazardous.

Avoid avalanche starting zones whenever you suspect instability. The characteristics of starting zones include (1) treeless or sparsely treed slopes lying at an angle greater than 30 degrees (a simple device to measure slope angles can be made from a plastic angle-protractor and a weighted string suspended from the center of the circle); (2) slopes that receive large amounts of snow, usually on the lee side of exposed ridges; and (3) gullies and bowls bounded at the top by horseshoe ridges or cliffs. Most starting zones are above the tree line, but some are below in open timber. Trees close enough together to be a nuisance to skiers are generally not active starting zones.

If you absolutely must cross an avalanche path or starting zone, take the following precautions to minimize the risk:

1. Move across the runout zone rather than the starting zone

because most avalanches that trap skiers are triggered by the victims themselves.

2. Keep to the flanks of starting zones when ascending or descending. This increases chances of skiing out of the slab area.

3. If you must cross a starting zone on the way to a ridge or saddle, enter as high as possible to keep the route short and to increase the chance of remaining on top of the snow should it let go. *Never* switchback up an avalanche path or a starting zone.

4. If possible, choose to cross a starting zone that empties into a flat, open runout zone rather than into a gully.

5. Avoid slopes that feed into cliffs, crevasses, and similar terrain hazards.

6. Avoid areas of unusually heavy snow accumulations, such as deposition areas on lee slopes.

7. Avoid areas of suspected wind slab.

8. If you suspect slab instability, choose a sunny slope rather than a cold, shaded slope, which would favor instability.

9. Cross dangerous slopes one at a time, trailing an avalanche cord. All members of the group should observe attentively. Do not stop in the middle of the slope. Other skiers should use the same track; additional tracks increase the probability of fracture and slab release. Just because one person has crossed safely does not mean the slope is safe. All too frequently, a slab is released by the second or third member of a party. If an avalanche catches and buries a skier, the others should concentrate on identifying and *marking* the point where the skier was visible.

Know the avalanche danger signs. In addition to terrain factors and current and precedent weather, there are other warning signs of potential avalanche danger:

1. Any avalanche observed nearby is an obvious danger sign. Snowpack conditions are usually more or less uniform over large areas, although there may be microclimatic variations that make a slope more or less stable than the average. In particular, slopes with the same aspect (that is, those facing the same direction) are likely to behave similarly, other factors being equal.

2. If a ski track propagates a fracture, instability is indicated. The deeper and more extensive the fracturing, the more unstable the snow.

3. Be alert for "noisy" snow. Fracture of the snow beneath the surface can be heard if not seen.

Be prepared to modify plans in the face of clear warning signs. Choose a new route or abandon the tour and retrace the route. An avalanche accident in the backcountry is much more likely to turn into an avalanche disaster than in a heavily populated ski area where help is minutes away. Finally, familiarize yourself with the following avalanche hazard checklist:

Slope	Angle between 30 degrees and 45 degrees
	Open or sparsely treed slopes
Topography	Obvious avalanche paths
	Lee slopes and bowls
	Lee side of bare ridges (cornice hazard)
	North-facing or shaded slopes
	Slopes above cliffs or gullies
Snow conditions	Snowpack depth greater than 2 feet
	Current sluffing or avalanching
	Ski traverse causes noisy or hollow sounds
	Ski traverse propagates fractures
	Mushy or waterlogged snow
Weather	Rapid accumulation rates
	Winds depositing snow in bowls and gullies
	Rising air temperatures or a change to rain during storm
	Thawing weather after a new snowfall
	Several weeks of cold weather following a major storm
	Springtime melt-freeze cycles
Time	During or within twenty-four hours of a storm
	Afternoon during warm spring weather

I have not attempted in this brief discussion of avalanche danger to detail rescue and survival techniques in the event of an avalanche disaster. A good source of information on winter skiing in general and avalanche hazards in particular is the Sierra Club totebook, *Wilderness Skiing,* by Tejada-Flores and Steck. Another must book for the backcountry skier is the *Avalanche Handbook* by Perla and Martinelli, published by the U.S. Department of Agriculture as Agriculture Handbook 489 and available from the superintendent of documents. It is a state-of-the-art compendium of what is known about snow and avalanches.

PART II

Regional Climatologies

Introduction

The chapters in this section describe the climate of various recreation regions of the North American continent in terms useful to those who find themselves outdoors in recreational pursuits. They answer three questions: What kind of weather am I likely to find in each region on a season-by-season basis? What is the best time of year to engage in my recreational activity? What hazards or discomforts am I likely to find?

Typical climatic averages are of little value in answering these questions. An average monthly rainfall of 4 inches means one thing to a backpacker if it is spread over the month in daily drizzle and light continuous rain and another if it occurs in six days of showers separated by periods of fine weather. I have attempted to develop meaningful statistics on the kind of weather that may be expected and on its distribution in time and space. I also consider the weather variability from one year to the next. Some regions have uniform conditions through the years; others vary considerably.

A compendium of conditions for all places recreationists use would fill a book several times this size. Some areas are seldom visited, and then primarily by local residents familiar with climatic conditions. So I have limited the number of areas to the most popular, those likely to attract visitors from far away. I have hiked in or visited all the regions included, and I know I would like to have had information on their climates when planning my trips

I have divided the continent into eight regions I think account for 90 percent of the backcountry recreation in the United States and Canada. They are the northern Appalachians, the southern Appalachians, the Great Lakes basin, the northern Rockies, the southern Rockies, the Cascades, the Olympic Peninsula, and the Sierra Nevada of California. If your favorite area is left out, do not be discouraged. That just indicates the need for a companion volume. Please let me know what areas should be included.

There is an inherent difficulty in producing a climatic description of outdoor recreation areas: they have been ignored by the meteorologists and climatologists. Most regularly reporting weather stations are in cities or, more commonly, at airports. Most published climatic maps are therefore maps of airport climatology. Few of us would want to hike on runways, even if we could. Climatological records, those taken by the vast network of cooperative observers, are somewhat better. But even these require someone living at the site; so they are generally located in valleys, on farms, or in other rural regions. The high mountains are a climatological wilderness.

A few mountain areas have regular reporting stations. One of the best, with a long history of high-quality observations, is on the summit of Mount Washington, well above the timberline in the White Mountains of New Hampshire. Other stations on mountain peaks or high passes have been occupied by observers for limited periods of time, usually as part of a research project on mountain weather. Berthoud Pass and a nearby summit in Colorado are examples.

Most of the data in the ensuing chapters were recorded at lower elevations in the mountain regions. I interpret these data in terms of the likely climate at higher elevations, especially above the timberline. Some caution in the interpretation and use of the data is necessary. Most of the mountain chains on the North American continent are oriented north-south and so extend over many degrees of latitude. Although the climatic latitudinal variation is less than the altitudinal variation (a rule of thumb is that 1 degree of latitude is climatologically equivalent to about a thousand feet of altitude), conditions in the north end of the Cascades differ from those in the south end. In defining the climatic regions, I had to compromise between splitting up a region into many little areas and trying to include too diverse a region. Fortunately, climatic controls are uniform over large areas; so it is possible to describe, for example, the climate of the thousand-mile-long Cascades by indicating something of the variations one can expect along the chain.

The scheme

The description of each region follows a pattern. The data are primarily from the published records of the U.S. National Weather Service and the Atmospheric Environment Service of Canada, supplemented by specialized observations and analyses where available and by my own observations and experience. The analyses are mine.

General climate

The section on general climate describes the major geographical features of the region that interact with the atmosphere's circulation pattern to produce that region's climatic pattern. I describe major storm tracks and indicate them on the regional map. This map also delineates the climatic region and locates the weather stations used for the analyses. This section gives a brief overview of the seasonal progression of climate as a background for the more detailed information contained in subsequent sections.

Solar data

Data on sunrise, solar noon (the clock time when the sun is due south of the observer), and sunset are given for one or two representative locations in each area. The length of civil twilight is also given. Civil twilight ends when the sun is six degrees below the horizon; it is normally dark enough then to prevent normal outdoor activities requiring natural light. However, there may be enough light to permit trail walking in open country for a period twice as long as the duration of civil twilight. Day length—the time between sunrise and sunset—is also included in the tables. This is the time available for hiking, skiing and route-finding.

Information is given on the variation of solar events such as sunrise and day length in each region. If you know your longitude (which can be read from topographic maps), it is easy to calculate the correction to the times of sunrise, solar noon and sunset. For each one-quarter degree you are east of the longitude of the location of the table, subtract one minute from the tabulated times; for each one-quarter degree west, add one minute.

Day length varies with both season and latitude. There is no simple way to calculate the correction, but the corrections for the winter and summer solstices for several locations in each region are given. Remember that day and night are of equal length everywhere at the spring and fall equinoxes, approximately March 21 and September 21. The shortest day is at the winter solstice, about December 21; and the longest day is at the summer solstice, June 21. Another point to remember is that these astronomical quantities vary most rapidly from day to day at the equinoxes and very slowly at the solstices.

Bioclimatic index

I described the rationale and use of the bioclimatic index in chapter 4. For each recreation region, I have chosen one, two, or sometimes three climatic stations representative of that region. Except for a few locations, these stations are below the timberline and are not representative of exposed high country. One can estimate the effect of altitude on the bioclimatic index by dropping the curves downward about 3 degrees for each thousand-foot gain in elevation (up to the timberline) and displacing the curve some distance to the right to account for the usual increase of precipitation with elevation.

Above the timberline, the curves should be displaced farther downward to account for the effect of wind speed on the windchill temperature. Below the timberline I assume air temperature and windchill temperature are equal because winds are generally below 3 mph in the trees.

The diagrams are plotted with a point for each month, using the monthly mean air temperature (or windchill temperature) and monthly mean precipitation. Points are connected in sequence; January is labeled "1" and December "12." Each bioclimatic type has a characteristic curve, or climogram. The curves thus present an overview of the bioclimate that characterizes each area.

The bioclimatic meaning of areas on this diagram and the use of the numbered zones in determining proper clothing are described in chapter 5.

Description of climate by season

The description of seasonal climates begins with a description of the summer climate. Temperature regime is presented for one or two locations in a diagram that gives monthly mean temperatures, average maximum and minimum temperatures, and extremes of record. The difference between the average maximum and the average minimum gives a clue to the typical daily range. Maritime climates have small daily ranges; inland climates have much larger ranges.

The curve connecting the monthly means also indicates the continentality of the climate. Regions near the coasts having a typical marine climate will have a flat annual curve; inland regions show a much larger variation.

Typical weather and sky conditions for each month of the year are combined in the next set of diagrams. Each month is represented by the number of days in it. (How much neater and easier to interpret they would be if we had months with equal numbers of days, with a five-day year-end holiday.) The shaded lower portion indicates the average number of days with cloudy daytime skies. A cloudy day is defined in climatological records as one in which the sky cover averages eight-tenths or more. The upper portion (with open circles) indicates essentially clear days, that is, days in which cloudiness averages two-tenths or less. The middle portion, with cumulus cloud symbols, indicates partly cloudy days, those in which cloudiness averages from three-tenths to seven-tenths.

Each month's typical precipitation pattern is indicated by symbols: a filled circle for a day with rain, a six-pointed snowflake for a day with snow, and a jagged thunderstorm symbol for a day with thunder. It is important to remember that the total number of symbols indicates the number of days each month with some kind of precipitation; however, a thunderstorm symbol does not necessarily indicate a day with rain. Climatic records do not distinguish between a thundershower at the observation point and a thunderstorm in the immediate vicinity that does not bring rain to the station. There is thus some ambiguity in the records; but in most places, a reported thunderstorm is accompanied by precipitation. Only in the Rocky Mountains, where many summer storms are "dry" lightning storms, may there be substantial ambiguity.

In the diagrams, a day with rain is one with 0.01 inch of precipitation or more. A snow day is one with 1 inch of snow or more. Because days with very light precipitation are included, the total number of days with precipitation of significance to the recreationist is exaggerated somewhat. The distribution between rain days and snow days in the winter is also distorted somewhat by inconsistencies in data reporting. For example, because 1 inch of snowfall is equal to approximately 0.1 inch of melted snow, and only snowfalls of 1 inch or greater are counted as snow, it is likely that some of the days listed as rain days in winter months are actually snow days with less than 1 inch of snow. But the published climatic data do not permit separation of these light snow days from rain days. Why are the published data not internally consistent? Ask the weather service!

Note that monthly and annual precipitation amounts are given in terms of "melted" precipitation whether or not the precipitation has occurred as rain or snow. Thus a foot of snow would be indicated as an inch of precipitation if it covered the ground to the depth of one inch when melted.

The third type of diagram included for every region shows the annual precipitation pattern, with monthly averages plotted as inches of rain or melted precipitation if it is in the form of snow. The numbers at the bottom of each month's column indicate the average total snowfall in inches of freshly fallen snow. A "T" indicates a trace, less than 0.1 inch. Annual totals are given, as well as the total snowfall (always greater than depth on the ground because of settling and melting).

Some of the chapters contain other diagrams when special analyses are appropriate and available. Most of these are self-explanatory; if they are not, a description is given in the text. The text describes climatic features not presented in the figures, such as snow depths, beginning and ending of seasons, peculiar local phenomena, and other useful information. To the extent possible, information on spatial and temporal climatic variability is presented. Remember that these are not forecasts but climatic statistics, primarily useful for planning purposes. When on the scene, keep your eye on the sky and sharpen your forecasting skills.

Additional information

The final section of each chapter gives references to particularly useful publications on the weather and climate of the region. Standard climatological publications may be obtained from the National Climatic Center in Asheville, North Carolina. Many libraries have the publications for the local area.

Figure 6-1. Northern Appalachians

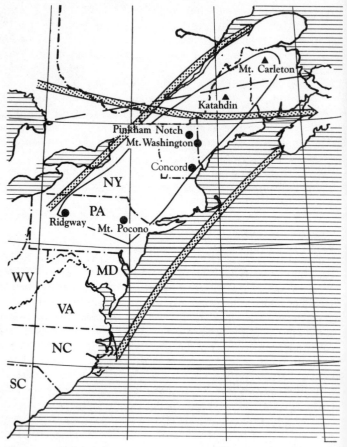

Arrows indicate winter storm tracks.

Chapter 6

The Northern Appalachians

Four hundred million years ago, giant crustal forces pushed up a range of mountains along the eastern edge of North America. Jagged peaks were thrust upward, forming a chain of superb alpine grandeur. But the slow and inevitable workings of wind, water, and ice ground them down and all that is left now is a series of low ridges and rounded mountains. They are low, that is, in comparison with the younger mountains of the earth. Tourists in Denver gazing westward toward the buttress of the Rocky Mountains are actually higher above sea level than if they were standing on the craggy summit of Katahdin in Maine.

But if ruggedness is measured in terms of the mountain environment, the Appalachians take their place among the most demanding in the world. Indeed, the weather on Mount Washington, highest in the Northeast, has been described as the worst in the world. Although that may be a slight exaggeration, the highest wind ever measured was recorded at the summit. On April 12, 1934, the summit observatory recorded a wind gust of 231 mph and a five-minute average wind speed of 188 mph.

The Appalachians stretch in a narrow band from northern Georgia northeastward to northern New Brunswick, roughly paralleling the Atlantic coastline. The highest peak, Mount Mitchell in North Carolina, 6,684 feet, is near the southern end. In the Northeast, Mount Washington is the tallest at 6,288 feet. At the very northern end, Mount Carleton, the highest peak in New Brunswick, is 2,690 feet.

In their northern half, the Appalachians are less of a chain than they are a series of isolated ranges: the Poconos and the Alleghenies in Pennsylvania, the Shawangunks in New Jersey, the Catskill Mountains and the Adirondacks* in New York, the Green Mountains in Vermont and their southern extension in Massachusetts, the White Mountains in New Hampshire, and the Longfellow Mountains in Maine, culminating in the magnificent Katahdin (5,267 feet). The Appalachian Trail traverses all the ranges except the Catskills and the Adirondacks in New York and the Allegheny Mountains in western Pennsylvania.

The northern half is distinguished from the southern portion of the range in another way: the mountains and valleys from Pennsylvania northward were heavily glaciated during the Pleistocene. The glaciation had a profound effect on the vegetation in eastern North America. Plant species were forced southward to find haven in the southern mountains. When the ice retreated, many of these species moved up along the mountain slopes as well as back northward to reclaim the barren land that had been scraped clean by the crushing ice-weight. Soil was built slowly on the granite bedrock and even today, some twelve thousand years after the ice sheet disappeared, the soil mantle is only a few inches thick on the mountainsides. Even so, many northern species and their genetic descendants occupy the highest mountaintops; the lowlands and the south have a varied, rich, almost tropical flora.

Today the northern Appalachians are a zone of ecological tension between the hardy spruce-fir forest that cloaks much of the Laurentian upland to the north and the temperate hardwood and pine forests of the south. In the highest elevations in the Adirondacks of New York, the Green Mountains of Vermont, the White Mountains of New Hampshire, and throughout northern Maine and New Brunswick, balsam fir and black spruce are the dominant species, interspersed with occasional showy stands

---

*The Adirondack Mountains are geologically part of the Laurentian Highlands. Climatically, however, they are closer to the Appalachians and so are included here.

of paper birch. At lower elevations, the forest is predominantly hardwood: the northern association of beech with an assortment of maples and birches. Farther south, white pine and hemlock mix to form a band stretching southeastward from the coast of Maine to central Pennsylvania. And stretching southward from this band lie extensive forests of oak and yellow poplar.

But the most interesting area remains to be described: the above-timberline tundra. Although the areas above timberline in the northern Appalachians comprise only a few square miles—the largest, in New Hampshire's Presidential Range, occupies about 8 square miles—they are one of the main attractions to hikers and backpackers. The climate and vegetation are characteristic of the lowland tundra 600 miles to the north, in Labrador. Every year thousands of hikers climb the summit of Mount Washington; thousands more ascend the mountain by cog railway or auto road.

Three other ridges in the White Mountains also have small alpine zones that attract hikers: Franconia Ridge, from Mount Lincoln to Mount Lafayette; part of the ridge between South Twin and Mount Guyot; and the summits of the Mahoosuc Range along the Maine-New Hampshire border. The Appalachian Trail traverses all the above-timberline areas in the White Mountains.

In addition, two other small alpine zones in the northeastern mountains attract hikers: Mount Marcy (5,344 feet) in the Adirondacks and Katahdin (5,267 feet) at the northern terminus of the Appalachian Trail in central Maine.

All these above-timberline zones share two things: hordes of hikers and severe weather. The summit of Mount Washington may not really have the worst weather in the world, but it is close to it. No region that has regular weather observations, including Antarctica, has worse weather. Timberline weather in the Appalachians demands respect and adequate preparation.

General climate

The climate of the northern end of the Appalachian chain is subarctic, even arctic, above timberline. At the southern tip, in Georgia, the climate borders on the subtropical. Although the transitions between the two are gradual and there are no distinct boundaries, a convenient climatic dividing line is the approximate southern boundary of the Pleistocene ice, in central Pennsylvania. From the Poconos north, the forests are indeed northern in appearance, and I have chosen the boundary somewhat arbitrarily to be where the Appalachian Trail crosses the

Susquehanna River at Harrisburg. I have also included the Adirondacks of New York, which climatically and vegetationally are part of the northern Appalachians and are properly included here.

The northeastern states lie in the zone of prevailing westerly winds and westerlies are predominant during periods of fair weather. In summer months, humid air from the Gulf of Mexico frequently streams up from the southwest, bringing periods of sultry weather. Occasionally, cool and damp air from the North Atlantic brings in cloudy, damp weather.

Low-pressure storms are frequent both summer and winter. Coastal storms tracking northeastward bring the heaviest rains in the spring, summer, and fall and the heaviest snowfalls in the winter. These are the classic northeasters. Storms with less precipitation arrive via another more or less parallel track down the Saint Lawrence Valley. Because this track brings southwesterly winds over New England, winter precipitation is more likely to be rain, especially in southern and coastal portions. These low-pressure storms often occur at intervals of three or four days, alternating with brief periods of fair weather. This rapid alternation of good and bad weather gives rise to an epigram usually attributed to Mark Twain, "If you don't like the weather, just wait a minute."

A major control over the weather of the northern Appalachians, especially in the White Mountains, is the position of the jet stream. When the axis of the jet stream lies athwart the White Mountains, extremely high northwest winds blow across the exposed ridges. There appears to be a kind of Bernoulli effect in which the normally high jet-stream winds accelerate to even greater velocities as they are squeezed over the mountains. At the summit of Mount Washington, the annual average is 35 mph, varying from a mean of 25 mph in July and August to 46 mph in January. Winds in excess of 100 mph are common. In 1977, for example, the fastest mile of wind passing the anemometer during June (the least windy month) was 89 mph. January, February, and March each had winds in excess of 138 mph. Hurricane-force winds (75 mph or greater) can be expected on half the days in the winter and on two to four days per month in the summer.

Just at the timberline (about 5,000 feet), wind speeds are somewhat less but still uncomfortable or even dangerous. Monthly averages range from about 15 mph in the summer to 30 mph in the winter. Maximum speeds have been measured at 50 mph in the summer and 100 mph in the winter. In the valleys, of course, speeds are much lower.

The annual temperature cycle in the Appalachians is mostly continental in character, despite the region's nearness to the Atlantic Ocean. The prevailing wind direction is from the west, which ensures this continentality. The annual range of mean temperature at Pinkham Notch, New Hampshire, is 47 degrees; at Mt. Pocono, Pennsylvania, it is only slightly less (43 degrees). The diurnal range is typically 20 degrees, although the summit of Mount Washington shows a smaller range, about 15 degrees.

About half the days throughout the year are overcast, somewhat more in the above-timberline areas. There is no marked annual pattern. About five days per month are clear, with a small increase in the number during the fall. The exposed ridges and summits are frequently immersed in cloud. To the hiker, this appears to be dense fog with visibility less than a quarter of a mile. On the summit of Mount Washington, 80 percent of all days, summer and winter, have dense fog for at least part of the day.

There is comparatively little north-south variation in annual precipitation amounts. In valley stations in the interior of New Brunswick, annual totals are around 40 inches. The 40-inch line snakes its way through Maine, New Hampshire, and Vermont, across New York State, and into northwestern Pennsylvania. The precipitation amount increases southward along the main axis of the mountains; at Mt. Pocono, for example, the annual average is 52 inches. Precipitation is spread uniformly throughout the year, with a slight tendency for a summer maximum. Summer rainfall is largely from showers, with four of these each month (more in the south) being thunderstorms.

The main control over precipitation amount is topographic. Although there are variations with season of the year and orientation of the slope with respect to storm direction, a rule of thumb is that the annual precipitation total increases 3 to 4 inches for each thousand feet of elevation. Snowfall also increases greatly with elevation: at Concord, New Hampshire (339 feet elevation), the annual total is 68 inches; at Pinkham Notch (2,000 feet elevation), the total is 168 inches; and at the summit of Mount Washington (6,288 feet), the annual amount is 247 inches. There is, of course, a north-south variation; areas in the Poconos receive about three-quarters of the snowfall that comparable locations farther north receive, despite the somewhat greater total annual precipitation.

Figure 6-2.   Bioclimatic index

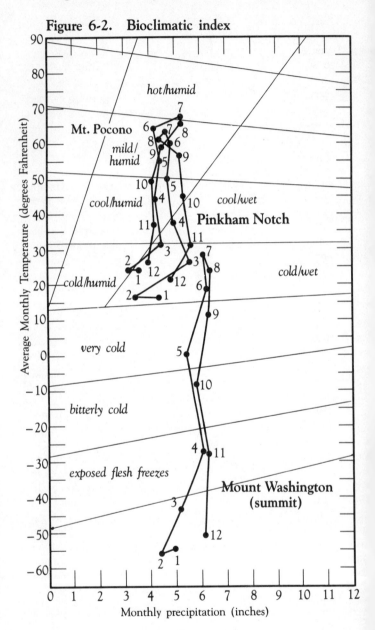

Thus the climate of the northern Appalachians can be characterized as a modified continental climate, with marked seasonal changes. Precipitation is largely the result of traveling low-pressure storms, but much summer rain comes from air-mass showers. Snowfall is substantial, although somewhat variable from year to year. Timberline is a major dividing line between a very severe climate above and the climatically more benign woodland below.

Solar data

Table 6-1 lists solar data for Mount Washington, New Hampshire. Day length varies from about 9 hours in mid-winter to 15½ hours in midsummer. Day length at Mount Carleton in midsummer is about one-half hour longer than at Mount Washington. For Mount Pocono, add about 20 minutes to the Mount Washington figures.

Bioclimatic index

Backpacker's bioclimate is shown in figure 6-2 for three typical locations. In the summer months, below-timberline areas are warm/humid; spring and fall are cool/humid; and winter months tend to be cold/wet. The more northerly reaches of the region tend to be somewhat cooler and moister, but the differences are not great.

Above timberline, however, low temperatures combined with high winds produce extremely hazardous conditions in all but the summer months. Even then the summer bioclimate can be characterized as cold/wet. From October through April, the bioclimate is bitterly cold. Windchill temperatures range from near 0°F to nearly 60°F below zero—extremely hazardous conditions for outdoor activity even for properly clothed persons. Of course, these are average windchill temperatures, based on monthly average air temperatures and mean wind speeds. Existing conditions will be better or worse, depending on the actual temperature and wind. But even the better conditions are severe and the above-timberline winter hiker must be prepared to endure extreme conditions. More important, above-timberline parties must be prepared to abandon planned goals and retreat below the timberline if conditions start to deteriorate.

Table 6-1.   Solar data for Mount Washington, New Hampshire

Date	Sunrise	Solar noon	Sunset	Day length hr:min	Twilight min
Jan 1	0721	1149	1616	8:55	33
Jan 16	0717	1155	1632	9:15	32
Feb 1	0704	1159	1654	9:50	31
Feb 16	0644	1159	1715	10:30	30
Mar 1	0622	1158	1733	11:12	29
Mar 16	0555	1154	1753	11:58	29
Apr 1	0526	1149	1812	12:47	29
Apr 16	0459	1145	1831	13:32	30
May 1	0439	1142	1849	14:14	32
May 16	0417	1142	1906	14:50	34
June 1	0404	1143	1922	15:18	35
June 16	0401	1146	1931	15:31	36
July 1	0405	1149	1933	15:27	36
July 16	0417	1151	1926	15:09	35
Aug 1	0433	1151	1910	14:37	33
Aug 16	0450	1149	1849	13:58	31
Sept 1	0509	1145	1822	13:13	30
Sept 16	0526	1140	1754	12:28	29
Oct 1	0543	1135	1726	11:43	29
Oct 16	0602	1131	1659	10:58	29
Nov 1	0623	1129	1635	10:12	30
Nov 16	0643	1130	1617	9:34	31
Dec 1	0701	1134	1605	9:06	32
Dec 16	0715	1141	1607	8:52	33

Note: Eastern Standard Time in hours and minutes on 24-hour clock. Add one hour during Daylight Time. Mt. Washington, New Hampshire is Lat. 44°16′N Long. 71°18′W.

Summer

When the snow goes, the black flies come. This usually happens by the end of May in the mountains of northern New England and by the end of March in the Poconos. By this time, the average daily temperature is above 50°F and summer has begun. By June at Pinkham Notch, minimum temperature will be below

Figure 6-3.   Temperature. Pinkham Notch, New Hampshire

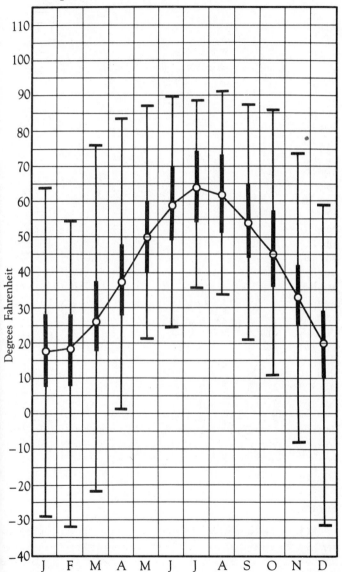

Figure 6-4. Temperature. Mount Washington, New Hampshire

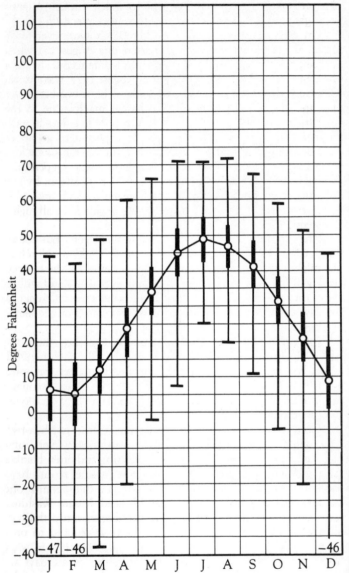

41°F on only five nights; half the time it will be above 48°F. Maximum temperature will be above 80°F on five days and above 70°F on fifteen. This pattern holds pretty much through August: five nights or so below the mid-forties, five days over 80°F.

On the summit of Mount Washington, summer never really comes, although I remember with pleasure one August day when the temperature very nearly equaled the all-time record. It was a warm and sunny 70°F; the record, set in August 1975, was a torrid 72°F. Such days on the summit are truly to be remembered, for they are very rare. More typically, summer maxima are in the mid-fifties with a brisk wind blowing. On only five days in each summer month can the temperature be expected to go above 60°F. Because of the nearly incessant wind, diurnal ranges are low, about 12 degrees in the summer. During June, night temperatures will fall below 29°F on five days and below 37°F on fifteen. By July and August, the comparable minima are up to 36°F and 42°F. This is small comfort when the wind blows at gale force all night, as it frequently does.

The farther south one goes, the warmer it gets. But conditions at Mt. Pocono are not as different from Pinkham Notch as one might expect. During the three summer months, the temperature will fall below the upper forties on five nights and will be above the mid-fifties half the time. There is even less difference in the daytime maximum temperature: Mt. Pocono is only about 2 or 3 degrees warmer than Pinkham.

Extremely high temperatures are uncommon throughout the region. The all-time maximum at Pinkham is 91°F and only 95°F at Mt. Pocono. The low at Pinkham for the three summer months is 25°F; at Mt. Pocono, 26°F. Above the timberline, of course, the extreme minima are lower, although the low record on Mount Washington in July is just 25°F.

Completely sunny days are rather rare in the northeastern mountains; cumulus clouds envelop the peaks nearly every summer day. Clouds start to build in mid-morning and often blanket the sky by early afternoon. Only about five days per month have daytime cloudiness of two-tenths or less and on the summits, one can count on only two such clear days. Indeed, Mount Washington is completely enveloped in cloud on more than two-thirds of all summer days. On these days, visibility is reduced to one-quarter mile or less. In the valleys, the situation is somewhat better. At Pinkham in August, only one-third of the days are predominantly cloudy, very similar to the southern portion of the region. About three days are foggy.

Figure 6-5.   Sky and weather. Mount Washington, New Hampshire

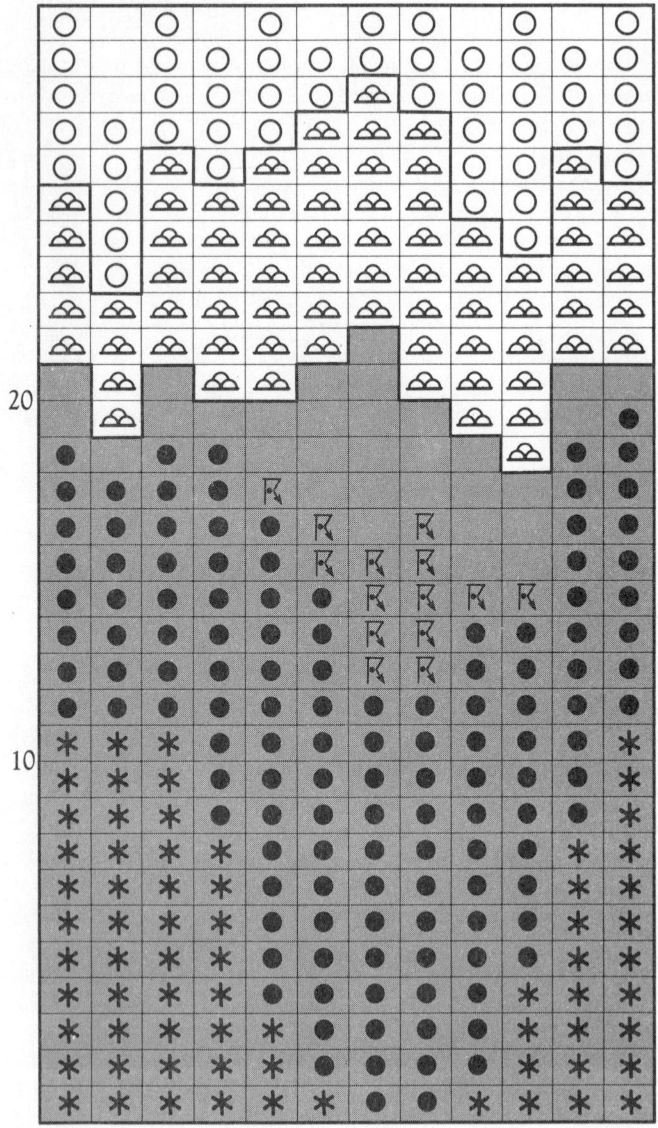

With summer clouds, unfortunately, comes rain. Most of this is from showers, which are over by nightfall. But traveling lows do bring southeasters even in the summer—about one each month. It will rain for two or three days straight.

The "thumpers," frequent thunderstorms, should concern the hiker. Above timberline in the Presidential Range, half the days have some kind of rain and a quarter of these will likely be thundershowers. Early buildups of towering cumulus clouds on moist and hazy days are a signal to plan to be off the ridges by early afternoon. Some of these will be crashingly heavy, complete with marble-sized hail. Below timberline, there are nearly as many showers and thundershowers, but they are of less concern. Farther south, only about a third of the days have rain; but because the air is more humid, more of these will build to thunderheads. This occurs about seven days per month.

Summers in the northeast mountains are wet: 4 or 5 inches per month at low elevations, up to 6 or 7 inches at high elevations and above timberline. Good rain protection is a must.

Table 6-2. Precipitation Frequency

| | Average Number of Days per Month with Precipitation Equal to or Greater than Indicated Amounts. | | | | | |
| | Mount Washington | | Pinkham Notch | | Mt. Pocono | |
Month	0.1″	0.5″	0.1″	0.5″	0.1″	0.5″
January	11	2	8	3	7	NA*
February	9	2	8	3	7	NA*
March	9	2	10	3	8	NA*
April	13	3	10	3	8	NA*
May	12	3	11	3	8	NA*
June	13	4	10	3	8	NA*
July	12	4	7	3	8	NA*
August	11	4	7	3	8	NA*
September	10	5	7	3	6	NA*
October	9	3	8	3	6	NA*
November	13	5	10	4	7	NA*
December	14	2	9	3	7	NA*

*Data not available.

Figure 6-6. Sky and weather. Pinkham Notch, New Hampshire

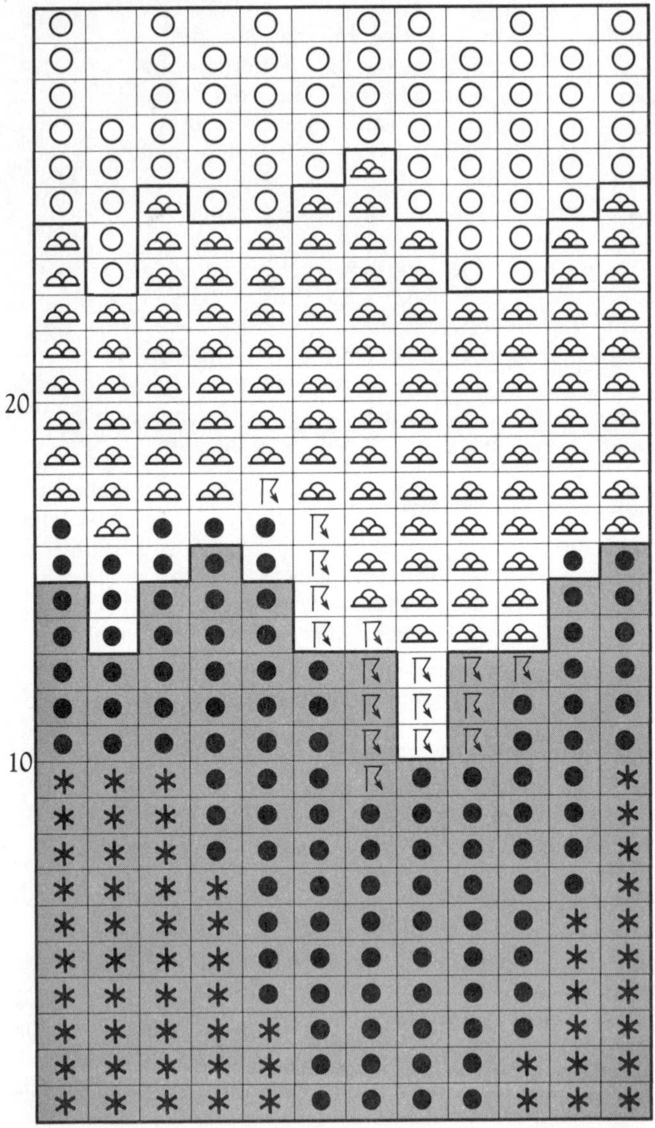

Autumn

The fall of the year is a delightful time in the Northeast. The Bermuda High starts to pull offshore and the flow of clear dry air from the north increases. Hardwood foliage begins its trip into senescence with a glorious burst of color. One always looks for the maples: were they really that red last year? There is probably nothing more delightful than to hike through the crisp air of an autumn day with newly fallen leaves crunching underfoot and the bright sun turning the dying leaves into patchworks of luminous reds and yellows; to be lulled to sleep by the rustle of leaves under the night wind; to see again the million stars in the sky, hidden all summer by a moist haze. A backpacker's heaven! If I sound predjudiced, I plead guilty. But if you find that combination of delights in the eastern woods, you will share my prejudice.

September and October are months to head for the backcountry. Temperatures cool rapidly; rainy and cloudy days are down; sunny days are up; and the bugs are gone. In the northern mountains, daytime highs will be above 66°F on half the days in September and above 58°F on half the days in October. Frosts become more frequent. Although only five days will have temperatures at or below freezing in September, more than half will be in October. Time to put your three-season sleeping bag to work.

Unfortunately, not all autumn days are clear and bright. About a third of the days will have some precipitation; and on one or two of these days, the rain will be accompanied by thunder and lightning. In the north country, some of the precipitation will be snow: a day or so in September, a couple of days in October.

Farther south, the scenario is similar but displaced in time by about a month. Snow usually does not come to the Poconos until November. By then, cloudiness and winter storms are on the increase; the bright days of early fall are finished.

Winter

In the north country, winter is in full swing by November. Swirling snow is as likely as rain in the low country; above 3,000 feet it is almost all snow. Snow amounts are variable. At Pinkham, as little as 2 inches and as much as 43 inches have fallen in November; but there is almost always a foot or more in December. The most snow that has ever fallen at Pinkham in December is 4 feet.

Figure 6-7. Sky and weather. Mt. Pocono, Pennsylvania

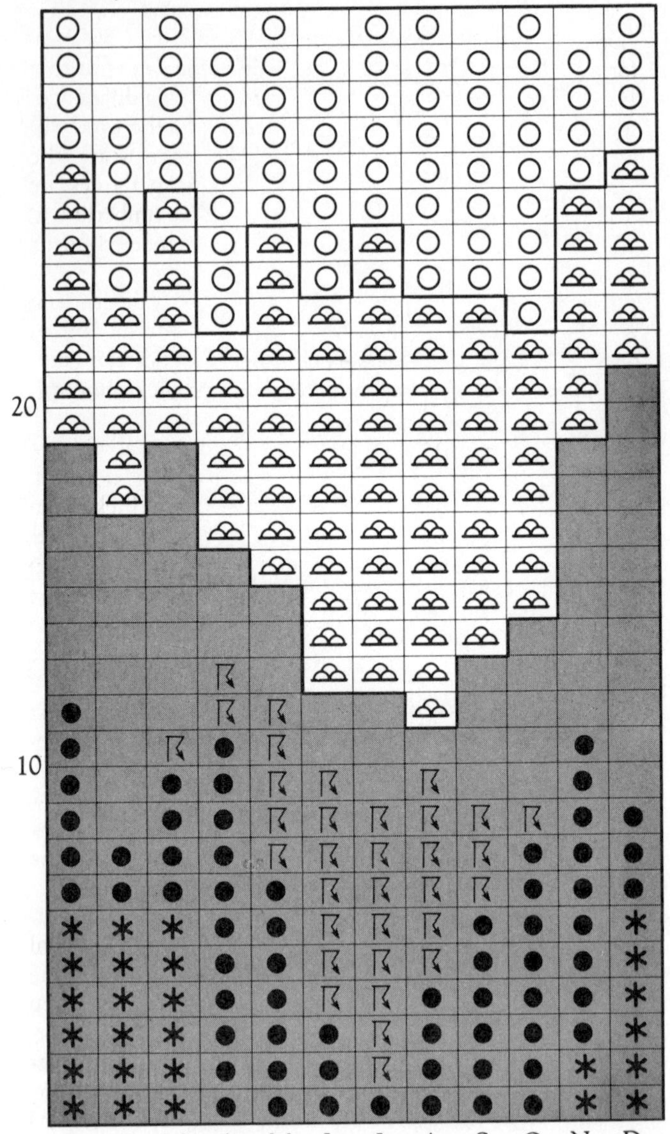

Temperatures are low in the north, averaging below freezing from mid-November until April. Minimum temperatures will be below 0°F on a third of the days in December, January, and February. Do not venture above timberline unless you are prepared for a true arctic experience. The temperatures are not that much lower than in the valleys, but wind makes a big difference. Wind speed on exposed ridges averages 50 mph, and the wind will be of hurricane force on half the days. To make matters worse, the winds are likely to be higher at night, when temperatures are lower, than during the daytime. So travel above the timberline in the winter is a demanding mountaineering experience requiring the utmost in skill and preparation.

One of the greatest winter hazards above the timberline is dense fog; the ridges and summits are in the clouds on more than three-quarters of the days. With snow on the ground, the landform may disappear and orientation become difficult or impossible: a true arctic whiteout.

A final winter hazard is snow avalanches, especially above timberline and in the open glacial cirques of the northern mountains. The most dangerous of these is probably Tuckerman Ravine on the eastern side of Mount Washington. Other places may avalanche more frequently, but none gets as much use by ice climbers and hardy skiers. The forest service maintains a regular avalanche patrol and posts closures at Pinkham Notch Camp at the beginning of the trail to Tuckerman's. These closures must be observed.

If I have scared you off the exposed peaks, let me take you below the timberline, where the weather is somewhat more benign. Winter in the northern Appalachians is generally suitable for all types of outdoor recreation. There are scores of ski areas— both downhill and cross-country—scattered throughout the mountains from the Poconos to New Brunswick and from the Atlantic coast to the Adirondacks.

Many of the areas have snow-making equipment, and temperatures are generally low enough for the production of artificial snow after mid-November in the north and mid-December in the south. The snow-making season can be considered to start when the average daily temperature falls below 28°F.

Most precipitation in the north country from November through April is in the form of snow. In the Poconos, rain may alternate with snow throughout the winter. Indeed, some years may find bare ground most of the winter in the southern end of

Figure 6-8.   Monthly precipitation

Mount Washington, New Hampshire

Annual: 76 inches   Snowfall: 247 inches

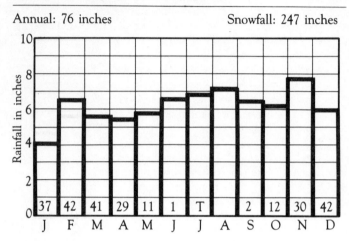

Ridgeway, Pennsylvania

Annual: 41 inches   Snowfall: 59 inches

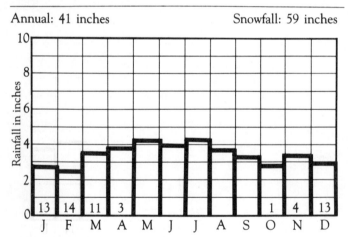

Figures at the base of the columns indicate inches of snowfall.

Figure 6-8. (*continued*)

Pinkham Notch, New Hampshire

Annual: 58 inches   Snowfall: 168 inches

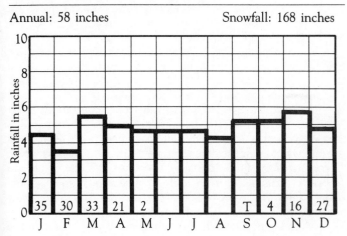

Mount Pocono, Pennsylvania

Annual: 52 inches   Snowfall: 55 inches

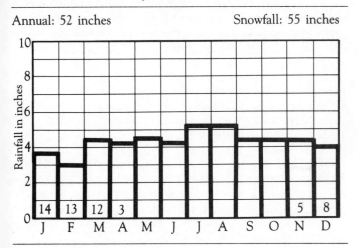

Figures at the base of the columns indicate inches of snowfall.

the region. In the north, the snowpack usually remains until May. But by this time, rain is as likely as snow and skiing becomes iffy. There is likely to be at least one thunderstorm in April above the timberline and one in the lower elevations by May. Farther south, thunderstorms start even earlier; winter usually ends in the Poconos by the end of March.

Winter skies are predominantly cloudy in the north country: over 50 percent are overcast and only about five days per month are clear. Fog is rare, occurring about one or two days per month. Conditions are similar throughout the region.

In midwinter, half the nights will have temperatures below 10°F; on one night in ten, the minimum will be below −10°F in the north country. Daytime maximum temperatures will go above freezing about one-quarter of the time and stay below 15°F about 10 percent of the time. In southern portions, half the nights will go below freezing; 10 percent will be below zero. In the daytime, temperatures will go above freezing on half the days but will stay below 20°F about 10 percent of the time.

November is usually the beginning of the snow season in the north country. The typical winter storm track is parallel to and just off the coastline. Counterclockwise circulation around the low center brings northeasterly winds and snow to the Appalachians. However, the exact path of the coastal storm is critical. A storm originating off the coast of the Carolinas may bring snow only to the southern portions of the region if the low center keeps well offshore. If it tracks northeastward toward the Gulf of Maine, northern New England and the Maritimes are in for heavy snow. If the storm moves northward, or if the storm has originated in the midwest and is moving down the Saint Lawrence Valley, winds over New England will be southerly and precipitation is likely to be rain rather than snow. So winds that stay southerly are bad news for Appalachian skiers; a good northeaster brings the snow.

The least amount of snow that has fallen in December at Pinkham in twenty-one years of record is 12 inches; the record low for January is 18 inches. Average amounts for these months are 27 and 35 inches, respectively. And the average stays at 30 inches or more through March. The snowpack reaches its usual maximum of 50 inches in March.

In the Poconos, the snowpack is more transitory. December through March, the probability that snow will cover the ground on any particular day is about one-third. The winters are variable:

some will be predominantly snowy; others will be "open," without a persistent snow cover.

In both areas, winter storms are generally of short duration: about two storms per month will last for two days; one will last for three days. There will normally be about three periods of three or more days without precipitation.

Wind speed in the timber is generally low, especially in the conifer forest, which maintains dense foliage throughout the year. In the more open deciduous forest, winter winds will be light (0–3 mph) more than half the time, moderate (4–12 mph) about one-fourth of the time, and greater than 25 mph about 5 percent of the time.

Spring

The snowpack starts to disappear in April in the north country and is usually gone by mid-May. Some snow is likely in May, but most of the precipitation is now in the form of rain. There may even be a thunderstorm in May; long-term records indicate an average of one. May is thus the transition month in the north country. April brings spring to the Poconos.

In April, 70 percent of the nights will have temperatures below freezing at Pinkham Notch; by May, the percentage will drop to fifteen. Maximum temperatures will also climb. 50 percent of April days will have daytime maximum temperatures above 47°F. By May, the 50-percent temperature is 62°F.

In the Poconos, only about one-third of April nights will be below freezing; but in May, the proportion drops to less than one night in ten. Fifty percent of April days will have afternoon temperatures above 56°F; 50 percent of May days will have temperatures above 66°F.

Spring also brings some decrease in daytime cloudiness. Nevertheless, about half the days are essentially overcast and only about six May days are sunny and clear. On the peaks and ridges, cloudy days increase to two-thirds and sunny days decrease to about four.

There is not much change in the total amount of precipitation or the number of rainy days in spring, although shower-type precipitation tends to displace the more persistent frontal precipitation as storm tracks move northward.

With the warming days of spring comes the scourge of the north country: black flies. These flying devils appear after the snow has melted and last well into the summer. Insect repellent, and lots of it, is essential. Trails may also be boggy at this time of year, but that is often a problem all summer long.

Summary

The northern Appalachians have generally good hiking weather from June through September or early October. At low elevations, normal summer hiking gear is adequate; but at high elevations and especially above the timberline, warm clothing, including wool hat and gloves, is essential. Rain is frequent and rain gear and tents are necessary for extended trips. Thunderstorms are common and precautions must be observed, especially above the timberline and on exposed ridges.

In the winter, climatic conditions are severe and require full winter gear, even in the southern portion of the region. Above timberline, winter weather is extreme; and full arctic equipment is mandatory. Snowstorms are frequent and sometimes prolonged, necessitating equipment for extended emergency bivouacking.

During spring and fall, conditions are extremely variable; and the backpacker must be prepared for rapid and occasionally violent deterioration of the weather. Close attention to weather signs and official forecasts is especially appropriate during these seasons. In the spring, insects may be a problem, especially in the north country.

Compared with other regions, the northern Appalachians have very favorable summer and early fall hiking conditions. Winters are snowy and challenging and provide generally excellent conditions for ski touring and ski mountaineering. At any season of the year, weather conditions can and often do change rapidly.

Additional information

Further information on the weather and climate of the northern Appalachians can be obtained in the following publications:

The Pocono Mountains of Pennsylvania. Climatic Summaries of Resort Areas. Climatography of the United States No. 21-36-1. U.S. Dept. of Commerce, Weather Bureau, 1969.

Saratoga Springs, New York. Climatic Summaries of Resort

Areas. Climatography of the United States No. 21-30-1. U.S. Dept of Commerce, Weather Bureau, 1963.

The above can be obtained from the National Climatic Data Center, NOAA, Federal Building, Asheville, NC 28801. They may also be available at state and local libraries.

Brooks, Charles F. "The Worst Weather in the World." *Appalachia* 6 (1940): 194–202.

Dethier, B. E., and Pack, A. B. "The Climate of Northern New York." Mimeographed. Ithaca, N.Y.: Cornell University, 1967.

Jackman, Albert H. *Handbook of Mt. Washington Environment.* Natick, Mass.: U.S. Department of the Army, Office of the Quartermaster General, Natick QM Research and Development Laboratory; 1953.

Ludlum, David. *The Country Journal New England Weather Book.* Boston: Houghton Mifflin, 1976.

Figure 7-1. Southern Appalachians

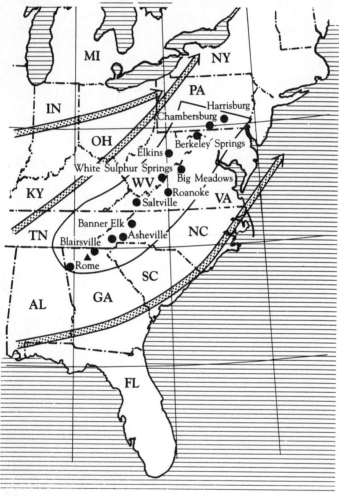

Arrows indicate winter storm tracks.

Chapter 7

The Southern Appalachians

Like folds in crumpled velvet, the southern Appalachians consist of a series of long ridges and intervening valleys stretching southwestward from Pennsylvania to northern Georgia. At the northern end, along the border between Virginia and West Virginia, at least three main chains can be identified: the Alleghenies on the west, the Blue Ridge on the eastern flank, and the Shenandoah Mountains in the middle. Farther north, the Appalachians lose their ridge-and-valley shape and become a series of isolated clumps of mountains. To the south, the ridges climb ever higher, culminating in Mount Mitchell (6,684 feet) near the southern end of the Blue Ridge.

Although geologically part of the great Appalachian chain that parallels the East Coast as far as Katahdin in central Maine, the southern Appalachians are distinct in nearly every other way. Central Pennsylvania makes a convenient divider. This marks the line of farthest southward advance of the vast continental glaciers that covered much of North America until just a few thousand years ago. The advancing glaciers pushed plant species ahead of them. As the ice retreated northward, so did the plants. Nevertheless, recolonization of the land laid bare by the retreating ice has been slow and uncertain. Relatively few of the plant species made it; consequently, the southern flora is much richer than the northern.

The southern edge of glaciation also makes an appropriate climatic divider. To the north, persistent snow cover holds the mountains in a wintry grip for several months; to the south, only the higher elevations develop a snowpack lasting for more than a few days after each snowfall. For the skier, this may be bad news; but for the hiker, it means the southern Appalachians offer year-round opportunities.

In addition to relatively mild winters, the southern Appalachians are climatically more benign than their northern neighbors in numerous ways. There is no true timberline in the southern mountains, although the ubiquitous grassy "bald" might be mistaken for alpine tundra. There is no counterpart to the strong rush of air the jet stream drives across the New England mountains with such ferocious velocities. Nevertheless, winter conditions can be severe, especially at the higher elevations; and the mountain traveler is well advised to watch for deteriorating conditions.

General climate

The main controls of southern Appalachian weather arise from the mountains' location near the southeastern edge of the continent. The Appalachians act as a substantial barrier to interchange of air masses between coastal regions and the interior. The eastern slopes are thus influenced greatly by the humid air masses from the Atlantic Ocean; the western slopes display a climate more typical of continental interiors.

In the summertime, the dominant influence is the Bermuda High, a large high-pressure area that sits over the South Atlantic and funnels warm humid air into the region. Pressure gradients are weak, however, and the air is often nearly stagnant. Sunny skies associated with the Bermuda High together with hydrocarbons produced by the trees that blanket the mountains conspire to produce the bluish haze that gave the Great Smoky Mountains their name. During the long summer season, the weather is typical airmass weather, except for the rare tropical storm or hurricane and the occasional passage of a frontal system.

In winter, the Gulf of Mexico is a breeding ground for the winter storms that dump quantities of rain and snow throughout the Northeast. Low-pressure systems that develop in the lower

Mississippi Valley usually track northeastward along the west side of the Appalachians (figure 7-1). These storms may be heavy rain producers (and occasionally heavy snow producers) in the western ridges because of the lifting of the southwesterly wind currents over the mountains.

A second major winter storm track is parallel to the Atlantic coastline but to the east of the mountains. These storms, on winds that are generally from the south and east, bring the heaviest precipitation amounts to the eastern slopes of the mountains.

After storm passage, the mountains are frequently invaded by cold, dry air from the north and west, flooding the region with crisp, clear Canadian air.

Annual precipitation amounts increase moderately from north to south. Harrisburg, Pennsylvania, and Berkeley Springs, West Virginia, have about 36–37 inches; Asheville, North Carolina, and Blairsville Experiment Station, Georgia, have 45 inches and 56 inches, respectively. Precipitation is distributed rather uniformly throughout the year but usually with a slight summer maximum. Most places have close to 4 inches per month. Summer precipitation is nearly all in the form of showers and thundershowers, whereas winter precipitation is more likely to be in the form of steady frontal rain or snow.

Snowfall shows both altitudinal and latitudinal variation. Winter totals range from 3 to 5 feet in the north (depending on elevation) to 0 to 2 feet at the southern end of the range.

Diurnal and annual temperature ranges are affected by the strong marine influence of the nearby Atlantic Ocean. Day-to-night variation is normally about 20 degrees every month of the year. The annual mean temperature curve is rather flat, with only 36 degrees separating the January and July means. The marine air also leads to generally cloudy conditions. At Elkins, West Virginia, for example, only three or four days per month are clear; the rest are cloudy or partly cloudy. Because of the usual presence of moist air in the summer and the more frequent invasion of Canadian air in the winter, there is typically a larger number of clear days in the fall and winter months than during July and August.

For the backpacker, the seasons may be divided into four somewhat unequal periods: a long summer, two short transitional seasons, and winter.

Solar data

Solar data for Big Meadows, VA, in Shenandoah National Park, are presented in Table 7-1. Day length is about 9½ hours in midwinter and nearly 15 hours in midsummer. Day length is about 12 minutes shorter at Harrisburg in midwinter; in midsummer, it's 12 minutes longer. At Asheville, North Carolina, day length is 16 minutes longer in summer, 16 minutes shorter in winter.

Table 7-1. Solar data for Big Meadows, Virginia

Date	Sunrise	Solar noon	Sunset	Day length hr:min	Twilight min
Jan 1	0731	1217	1703	9:32	29
Jan 16	0730	1223	1717	9:47	29
Feb 1	0720	1227	1735	10:15	28
Feb 16	0704	1228	1752	10:48	27
Mar 1	0645	1226	1807	11:22	26
Mar 16	0623	1222	1822	11:59	26
Apr 1	0558	1217	1837	12:39	27
Apr 16	0536	1213	1851	13:16	27
May 1	0516	1211	1906	13:50	28
May 16	0501	1210	1919	14:19	30
June 1	0451	1212	1932	14:41	31
June 16	0449	1214	1940	14:51	31
July 1	0453	1218	1942	14:48	32
July 16	0503	1220	1937	14:34	31
Aug 1	0516	1220	1924	14:08	29
Aug 16	0529	1218	1907	13:37	28
Sept 1	0543	1214	1844	13:00	27
Sept 16	0557	1208	1820	12:24	26
Oct 1	0610	1203	1757	11:47	26
Oct 16	0624	1159	1734	11:10	27
Nov 1	0641	1157	1714	10:33	27
Nov 16	0657	1159	1700	10:03	28
Dec 1	0713	1203	1653	9:40	29
Dec 16	0725	1210	1654	9:29	30

Note: Eastern Standard Time in hours and minutes on 24-hour clock. Add one hour during Daylight Time. Big Meadows, Virginia is Lat. 38°31′N Long. 78°26′W.

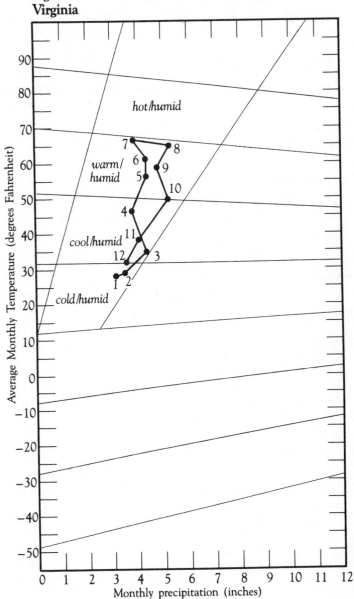

Figure 7-2. Bioclimatic index. Big Meadows, Virginia

Bioclimatic index

The suitability of the region's climate for outdoor activity can be determined from the bioclimatic index calculated for Big Meadows, the headquarters of Shenandoah National Park in the Shenandoah Mountains of Virginia. (The bioclimatic index calculated for Blairsville, Georgia, at the southern end of the region shows only slightly warmer and more humid conditions. The Big Meadows index can therefore be taken as representative of the entire high-mountain region.)

December, January, and February are only marginally characterized as cold/humid. However, backpackers should be prepared for brief periods of cold weather. The long summer, May through September, is warm/humid, verging on hot/humid, especially farther south and at lower elevations. The transition months of March–April and October–November are cool/humid. The climate is appropriate for backpacking throughout the spring, summer, fall, and even during the winter, with exceptions noted in the following sections. Cross-country skiing is marginal and downhill ski areas must rely on snow-making equipment for adequate slope coverage. Even then, winters are sometimes too mild to permit adequate amounts of artificial snow, except in the northern end of the region.

Summer

Summer begins early in the southern Appalachians. By April, daytime temperatures rise above 50°F on nearly all days even at high elevations. On half of May days, at intermediate elevations, maximum temperatures will be in the mid-seventies; and on five days, the maximum will be in the mid-eighties. At higher elevations near the crest, maximum temperatures will be 8 to 10 degrees lower. Freezing temperatures will occur on only one or two nights during the month. July is the warmest month, with the mean temperature throughout the mountains being very close to 74°F. In terms of temperatures, August and September are nearly symmetrical with June and May.

Day-to-night temperature variation is comfortably large during the summer season; it almost always cools down enough at night to permit a good night's sleep. On only one day out of eight is the minimum temperature less than 19 degrees lower than the maximum temperature. Extremely hot or cold temperatures are rare. The record high at Elkins was 97°F in September 1953; at

Figure 7-3. Maximum temperatures

Maximum temperatures will be higher than indicated on 5 days each month

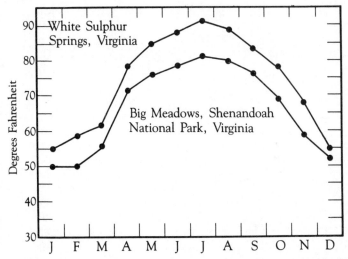

Maximum temperatures will be higher than indicated on 15 days each month

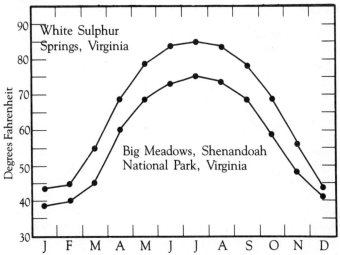

Asheville, it was 96°F in June 1969; and at Big Meadows, high in the Shenandoah Mountains, the all-time maximum reached only 90°F (September 1954). Record minimum temperatures for the same locations were −22°F, −7°F, and −14°F, respectively.

Although rain is frequent in the summer—about one day in four—most of the rain comes in short showers. On only two or three days each month will the rainfall amount be greater than 0.5 inch. Two periods of three or more consecutive days each month can be expected to be rainy. Rainfall amounts and frequency increase somewhat from north to south. In the Great Smokies, the increase over the northern portion will be about 25 percent.

Table 7-2.　Percentage Frequency of Daily Temperature Ranges, by Seasons. Berkeley Springs, West Virginia.

Range (Max– Min.) °F	Winter (Nov.– Mar.) %	Spring (Apr.– May) %	Summer (June– Aug) %	Autumn (Sep.– Oct.) %
0–9	9	3	0	2
10–18	28	21	13	13
19–27	36	30	32	25
28–36	18	27	34	29
37–45	8	14	18	26
46–54	1	5	3	4
>54	0	0	0	1

With summer showers comes the increased likelihood of thunderstorms. Six or seven days can be expected to produce "thumpers" in the northern portion of the region, increasing to ten to thirteen days in the southern end of the range, during the peak of the summer season. Summer days typically develop cumulus clouds early on, with some of these turning into shower clouds. Clear days are rather rare in the summer, with only 15 to 20 percent being so categorized in July in Asheville, North Carolina. Cloudy and partly cloudy skies rule the mountains. However, summer months can be expected to have two or three periods of dry spells lasting four or more days.

Shower and thunderstorm activity tapers off in late summer. September normally will have only three electrical storms, and

Figure 7-4. Temperature. Asheville, North Carolina

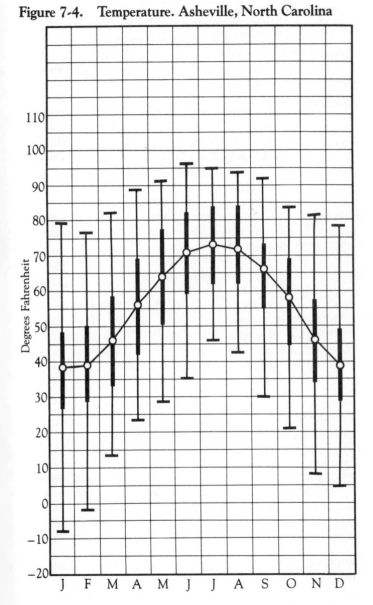

these generally come during the first half of the month. An all-day rain can be expected only about once during September.

Summer is thus a good time for hiking in the southern Appalachians. Although rainshowers are frequent, they are generally of short duration. Days can be sultry, with high temperatures and humidities; but nights are generally cool.

Autumn

October and November are transition months. Summer is over, but winter has not yet begun. Temperatures moderate considerably; maximum temperatures in the high seventies are rather rare, especially in October. However, maximum temperatures will be 70°F or above on half the days in October and above 56°F on half of the days in November at intermediate elevations. At higher elevations, maximum temperatures will be some 10 degrees lower. Nights become increasingly chilly, with minimum temperatures below freezing occurring on about eight nights in October and over half the nights in November. The spread between daytime maximum and nighttime minimum temperatures is even greater than in the summer, largely because of a larger proportion of clear days and nights. In both months, clear skies occur on nearly 40 percent of all days. Fall is a time of frequent invasions of clear, cold air from the north, producing the bright, clear weather so characteristic of autumn in the East.

As a result of the persistence of this relatively clear weather, precipitation amounts are less than in the summer. The amounts range from about 2.5 inches per month in the north to nearly 4 inches in the south.

Thunderstorms fall off drastically in autumn, with only one or two occurring each month. An October snowfall is a rarity, but November typically sees a couple of inches, except in the southern tip of the Appalachians.

October and November are thus ideal hiking months in the southern Appalachians. Weather is often crisp and clear, with mild days and cool to cold nights. An added advantage of fall hiking in the deciduous forest of the East is the brilliant fall foliage. For many backpackers, it is the preferred season.

Winter

Winters are short and relatively mild in the southern Appalachians; although at higher altitudes (above about 3,000 feet), conditions can be similar to those in the Appalachians many

Figure 7-5. Sky and weather. Elkins, West Virginia

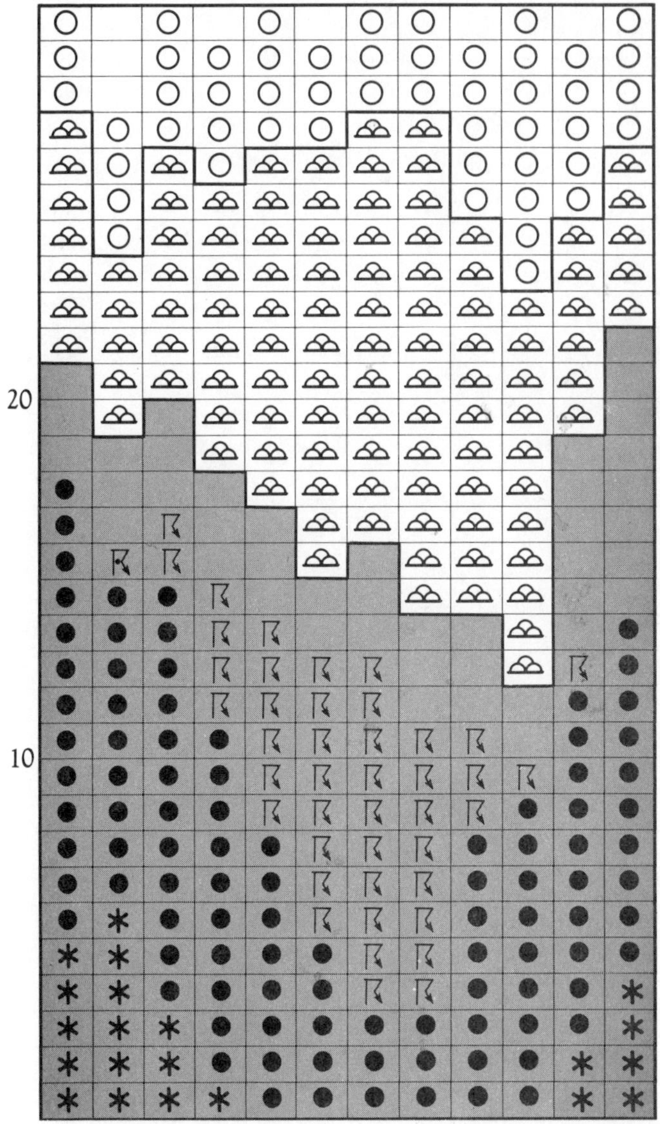

Figure 7-5. *(continued)* **Asheville, North Carolina**

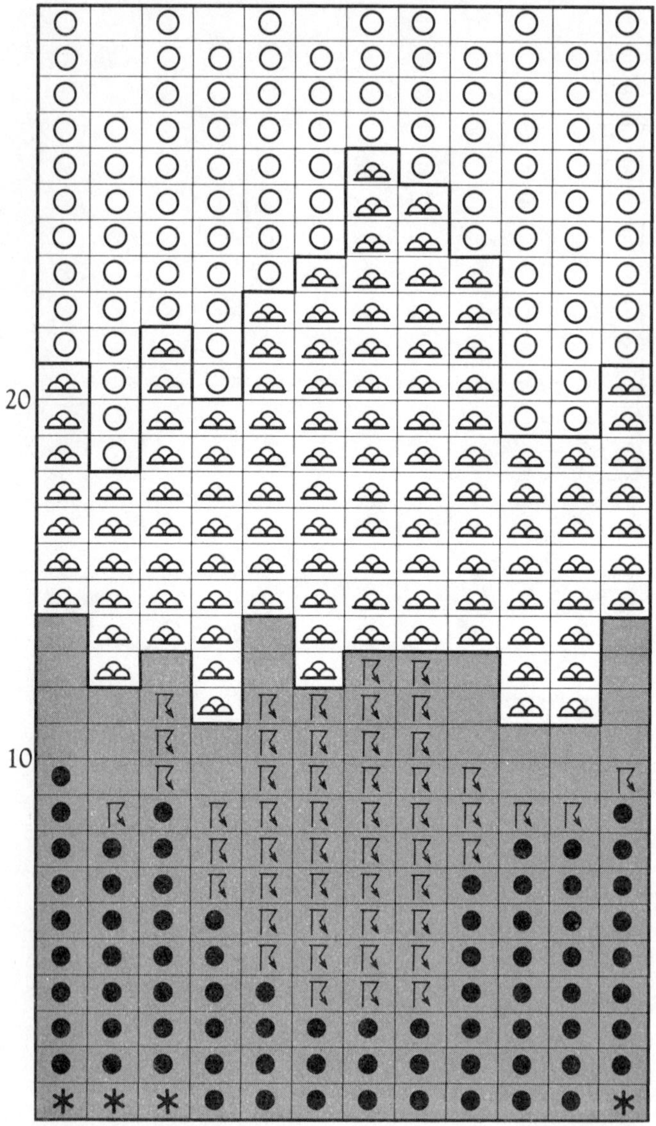

Figure 7-6. Monthly precipitation

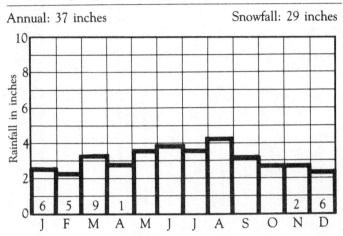

Berkeley Springs, West Virginia

Annual: 37 inches   Snowfall: 29 inches

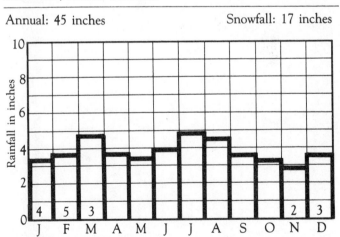

Asheville, North Carolina

Annual: 45 inches   Snowfall: 17 inches

Figures at the base of the columns indicate inches of snowfall.

Figure 7-6. *(continued)*

Shenandoah National Park, Virginia

Annual: 52 inches   Snowfall: 36 inches

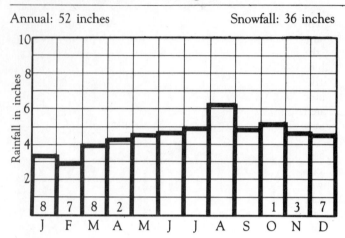

Blairsville, Georgia

Annual: 56 inches   Snowfall: 6 inches

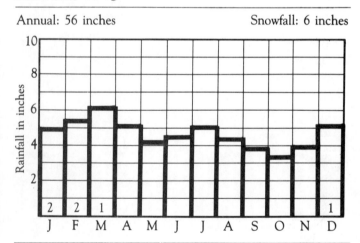

Figures at the base of the columns indicate inches of snowfall

hundreds of miles to the north. The mean January temperature at Harrisburg, Pennsylvania (338 feet altitude), is 30°F. At Asheville, North Carolina, 300 miles farther south but nearly 2,000 feet higher, the January mean is 38°F, 8 degrees warmer. More typical of the higher elevations is Banner Elk, North Carolina (3,750 feet); the January mean is 32°F. This verifies the rule of thumb that mean temperature decreases about 3 degrees for each thousand feet of elevation.

Although daytime maximum temperatures may be mild (maximum temperatures at White Sulphur Springs, West Virginia, are higher than 43°F on half of all winter days), nocturnal minimums below freezing are reached on three-quarters of the nights.

Most winter precipitation is in the form of rain; but snow is not uncommon, even in the mountains farthest south. At Big Meadows in Shenandoah National Park, average winter snowfall is 48 inches. This amount decreases southward, diminishing to 6 inches at Blairsville in northern Georgia. The ridges exert a strong orographic control, however, and higher elevations receive proportionally more snow. For example, at Asheville, winter snowfall averages 17 inches; Banner Elk, a few miles distant and 1,600 feet higher, receives an average winter snowfall of 45 inches, two and a half times as much. However, just to show that generalizations in weather and climate are always somewhat dangerous, consider Pickens, West Virginia. Here, at an elevation of 2,700 feet, topography and exposure have conspired to produce an average annual snowfall of 145 inches.

Thunderstorms are rare in the winter months, although the southern end of the region may experience one or two every winter.

Winter months are not the most propitious for backpackers in the southern Appalachians. In most of the area, there is too little snow for skiing and too much snow for good hiking.

Spring

The saying "March comes in like a lion and goes out like a lamb" has considerable substance to it in the southern Appalachians. Northward incursions of very moist air from the Gulf of Mexico clash with winter's last southward push of cold air. The frequent result is a period of blustery and rainy, sometimes snowy weather during the first half of March. Indeed, some of winter's heaviest snowfalls occur during this period (the blizzard of 1888 started on March 11). But winter's grip loosens during the latter half of the month and temperatures ameliorate markedly. At White Sulphur Springs, West Virginia, half of all March days have maximum temperatures higher than 55°F; in April, the median temperature is up to 68°F. Even at higher elevations, as at Big Meadows in Shenandoah National Park, half of April days are above 60°F. Spring arrives rapidly as the atmospheric circulation patterns swing to a persistent southerly flow. The maximum temperature will be above the mid-seventies on five April days, a sure sign that warm air is flooding northward from subtropical areas.

Cold nights continue, however, especially in March; and even in April, about a quarter of the nights will have temperatures below freezing.

Precipitation amounts during March are usually somewhat greater than the months preceding and following, although the differences are not great. Only three or four days each month will have more than 0.5 inch of precipitation; about eight will have rainfall of 0.1 inch or more. Dry spells of four or more consecutive days can be expected twice in each of the two transition months; expect about the same number of rainy spells of three or more successive days.

With the transition to more summerlike weather, thunderstorms also increase. March may have only one or two storms, but April may have three or four.

Because of the profusion of wildflowers and especially the rhododendron during spring, March and April are popular hiking months. Backpackers must be prepared for changeable weather, however; summerlike days can be followed by blustery and cold days with startling rapidity.

Summary

The southern Appalachians have good hiking weather from April through October. May and June are especially popular for their spectacular displays of flowering rhododendron. Normal summer hiking gear is appropriate but the frequent rainfall indicates at least a poncho. A light or medium weight sleeping bag is sufficient. A light-weight tent, perhaps a tube tent, is essential unless you are planning to utilize the trail shelters that are common in the southern Appalachians. Thunderstorms are common, especially in mid-summer.

Hiking is still possible most of the rest of the year, especially in the southern half of the region. Winter storms occasionally bring snow. Winter clothing, a medium weight sleeping bag and a good tent are necessary for extended trips. In the northern half of the region at the higher elevations, severe winter storms with snow deep enough to make walking difficult must be prepared for.

In the early spring months of March and April, mosquitoes, black flies and deer flies can be a problem. Ticks are most common during these months, too.

Additional information

Several publications of the National Weather Service (available from the National Climatic Center, Federal Building, Asheville, NC 28801) contain climatic summaries of interest to outdoor recreationists. They are:

Shenandoah National Park, Virginia. Climatic Summaries of Resort Areas, Climatography of the United States No. 21-44-1. 1969.

Berkeley Springs, West Virginia. Climatic Summaries of Resort Areas, Climatography of the United States No. 21-46-1. 1963.

White Sulphur Springs, West Virginia. Climatic Summaries of Resort Areas, Climatography of the United States No. 21-46-2. 1963.

Figure 8-1. Great Lakes Basin

Arrows indicate winter storm tracks.

Chapter 8

Great Lakes Basin

The great Laurentian Shield, worn down by countless eons of wind and water, suffered a final degradation when the Laurentide ice sheets of the Pleistocene scoured it clean and deposited the debris on what is now the northern tier of the United States. Typical glacial features abound in this area: drift piled up in terminal moraines, drumlins, eskers, and kettles. In some places, great blocks of ice were buried by the glacial debris as the main ice mass eroded and melted. When these gigantic ice cubes melted, deep kettles formed that now contain sparkling cold lakes. Some fine examples of these ice-block lakes can be found in the northern end of Michigan's Lower Peninsula.

Drainage in this periglacial region is often chaotic: lakes with no outlet or perhaps two outlets draining in two directions; low hills of drift oriented every which way, depending on the vagaries of the melting ice. Placid, often swampy streams are interrupted by stretches of white water.

As the ice sheet melted back toward the north, a vast lake formed along the southern edge, for the land here sloped gently north. When the ice was finally gone, a giant basin remained, forming the world's greatest expanse of inland fresh water: the Great Lakes. These inland seas are unique in several respects. Most lakes are formed by some interruption of the normal drainage pattern of a watershed, a landslide or a fault damming a stream or river. Such lakes occupy a small portion of the entire drainage. But the Great Lakes occupy nearly a third of their basin (95,000 square miles out of a total of 295,000 square miles). The divide between the basin and surrounding drainage is rarely more than 100 miles from the lake shores; and at one point on the southwest shore of Lake Michigan, it is a scant 2 miles away.

The vegetation in the basin is transitional between the oak woodlands to the south and the vast spruce-fir forests of the north. South of Lakes Ontario and Erie lies a band of hardwoods—oak, yellow poplar, ash, and occasional hickories. Farther west, in northern Michigan and northern Wisconsin, beech-birch-maple stands alternate with patches of aspen and birch. Occasional stands of spruce and fir are found here, but they are more common in northern Minnesota. In the Boundary Waters Canoe Area, red pine is found, often in sizable pure stands.

North of the lakes, spruce and fir predominate; but the pines are also found, as well as the picturesque white birch and quaking aspen. To the east, extensive stands of second-growth white and red pine occupy areas that have been disturbed by logging or fire. Aspen also covers large areas that have been recently burned. Hardwood forests of birch, beech, and maple are found on the relatively rich loamy soils of the Algonquin Highlands.

The Great Lakes basin is a mosaic of hardwoods and conifers interspersed with farmlands, of lakes by the thousands and streams by the tens of thousands, of vast areas of wilderness, and of provincial parks and national forests by the score. It is an area where much of the backcountry recreation is pursued by canoe in the summer and by ski or snowmobile in the winter. Mountain climbing in not one of the attractions: the highest point in the basin (outside of the Adirondacks, which form the eastern divide) is Ishpatina Ridge, north of Sudbury, Ontario, with an elevation of 2,274 feet. Tip Top Mountain (2,099 feet) in Pukaskwa Park on the north shore of Lake Superior is only 1,500 feet above the lake level.

The region described in this chapter is essentially coterminous with the Great Lakes basin. However, a few immediately adjacent areas, such as Algonquin Park and the Boundary Waters-Quetico area, have been included.

General climate

The Great Lakes basin is far removed from the ocean, and the climate might be expected to be predominantly continental in character—large diurnal and annual temperature ranges and marked seasonal contrasts. And so it is; but the huge inland lakes exert a profound influence on the climate near their shores, especially on the downwind side. Milwaukee and Muskegon are at the same latitude, 80 miles apart on opposite shores of Lake Michigan. Milwaukee is on the western shore and subject to the general movement of air from the interior; the same air must pass over the lake before reaching Muskegon. In January, the mean temperature is 3 degrees warmer in Muskegon; in July, it is 1.5 degrees cooler. The annual precipitation is 1.5 inches greater on the eastern shore of the lake; winter daytime cloudiness is greater; humidity is higher. The lake thus exerts a moderating influence on temperature but increases moisture and precipitation.

The other major modifying influence on the region's climate is latitude. Armstrong, near the north end of Lake Nipigon, is 400 miles—nearly 6 degrees of latitude—north of Houghton Lake, Michigan. Both have a yearly precipitation of about 29 inches. But the mean annual temperature at Armstrong is 12 degrees lower than at Houghton Lake. Most of this difference occurs in the winter: the daily mean temperature in January at Armstrong is 23 degrees lower; in July, the difference is only 5 degrees.

Unlike the other recreation regions described in this book, great difference in elevation is not a major influence, although topography may exert considerable microclimatic control. Most of the terrain north of the lakes is typical of Precambrian topography, with rocky upland of low relief. South of the lakes, the terrain is even gentler.

Thus steep environmental gradients are confined to the first few miles inland from the shorelines of the large lakes. There is considerable climatic homogeneity in the basin.

The Great Lakes basin lies between two major atmospheric airmass areas: the Canadian interior, source of cold, dry air; and the Gulf of Mexico, source of warm, moist air. The boundary between cold polar air and moist tropical air can frequently be found in the region. Because of this, storms often originate here and move eastward or northeastward along the polar front. These storms account for about 20 percent of those found in the region. Most of the winter storms originating outside the region come from the southwest (38 percent), the west (27 percent), or the northwest (9 percent).

Storms occur rather frequently and pass through rapidly, contributing to great day-to-day variety in the weather. In January, an average of eight storms originate in or pass through the region, about one every four days. In the summer, storm frequency is about half the January average.

Winter storms are frequently followed by short periods of clear, dry, and very cold weather from the arctic. The dominant winter air mass, however, is of Pacific origin. Although these air masses lose most of their moisture in crossing the coastal mountains, they pick up some moisture from the plains and arrive somewhat moister than the arctic air masses. Rarely does the moist air from the Gulf of Mexico invade the region at ground level; it tends to override the cold air and contributes to winter cloudiness and precipitation.

Annual precipitation averages between 26 and 52 inches and is distributed rather uniformly throughout the year. About a quarter of the total occurs as snowfall. The highest amounts are found in the extreme eastern portion of the basin, just east of Lake Ontario, and are due to the lake's effect of increasing the moisture content of the air passing over it. This lake effect also produces annual amounts on the order of 40 inches just east of Lakes Superior and Huron. Elsewhere amounts are generally between 30 and 34 inches.

There is a marked lack of seasonality in the precipitation pattern. Locations with a strong continental influence, such as the Boundary Waters Canoe Area, have a marked precipitation maximum in the summer. Farther east, as at Houghton Lake, the summer maximum is less pronounced. Immediately downwind of the Great Lakes the annual pattern is more uniform. Winter snowfall is increased by the lake effect: cold Canadian air passing over the relatively warm water becomes moister and more unstable. The opposite effect occurs in the summer: warm air passing over the cool water stabilizes the lower atmosphere and results in depressing convective storms along the downwind shores. Just east of Lake Superior these effects are sufficient to reverse the typical pattern. At Wiarton, on the Bruce Peninsula jutting into Lake Huron, winter precipitation is a third more than summer precipitation.

In the summer, the dominant circulation is still from the west, with Pacific air masses present about 30 percent of the time, somewhat more frequently north of the lakes, somewhat less to the south. This air is relatively cool and dry and brings bright skies and clear air. Hot, humid air from the gulf is present about a third of the time south of the lakes but pushes north of Lake

Superior only about 10 percent of the time. Only occasionally does hot, dry air from the Southwest invade the region.

Spring and fall are periods of transition. Frontal systems move rapidly through the region and produce frequent, widespread showers and cloudy skies.

Near the shores of the Great Lakes, lake breezes predominate during periods of quiescent weather. In the summer, the daytime breeze is from lake to shore; in winter, the direction is reversed because the water is warmer than the land. During stormy conditions, wind direction and strength are controlled by the position, intensity, and direction of the storm. In the winter, the winds have a predominant westerly component.

Solar data

Data are presented for two locations: Algonquin Park in the eastern portion of the area (Table 8-1); and the Boundary Waters Canoe Area in the far western portion (Table 8-2). At both locations, summer day length is nearly 16 hours; winter day length is less than nine.

Most of the Great Lakes Basin is in the Eastern Standard Time zone. However, the Boundary Waters Canoe Area is in the Central Standard Time zone. Table 8-2 can be used for Thunder Bay, about two degrees east of the BWCA but in the Eastern zone, by adding 52 minutes to the tabulated times of sunrise, solar noon and sunset. Day length will be about the same at Thunder Bay and the BWCA.

Bioclimatic index

Because of the slight summer maximum in precipitation, the bioclimatic index south of the Great Lakes remains in the humid zone all year, ranging from cold/humid in the winter to mild/humid in the summer. The climogram for Houghton Lake is typical of this pattern. North of the lakes, the pattern is similar but displaced downward. The climogram of Atikokan, north of the Boundary Waters Canoe Area indicates that December, January, and February have a very cold (zone II on Figure 8-2) bioclimate. June through September is very similar to Houghton Lake but a few degrees cooler.

Typical of continental climates, large temperature extremes occur with some regularity, pushing the winter bioclimate into the bitterly cold or dangerously cold zones. The climate of the

Table 8-1.   Solar data for Algonquin Park, Ontario

Date	Sunrise	Solar noon	Sunset	Day length hr:min	Twilight min
Jan 1	0755	1217	1639	8:44	34
Jan 16	0751	1223	1655	9:05	33
Feb 1	0736	1227	1718	9:42	32
Feb 16	0715	1227	1740	10:25	31
Mar 1	0651	1226	1800	11:09	30
Mar 16	0623	1222	1821	11:57	30
Apr 1	0552	1217	1842	12:49	30
Apr 16	0525	1213	1901	13:37	32
May 1	0500	1210	1921	14:21	33
May 16	0440	1210	1939	15:00	36
June 1	0426	1211	1956	15:30	38
June 16	0423	1214	2006	15:43	39
July 1	0427	1217	2007	15:40	38
July 16	0439	1219	1959	15:20	37
Aug 1	0457	1220	1942	14:46	35
Aug 16	0515	1217	1920	14:05	32
Sept 1	0535	1213	1852	13:17	31
Sept 16	0553	1208	1823	12:29	30
Oct 1	0612	1203	1753	11:41	30
Oct 16	0632	1159	1726	10:54	30
Nov 1	0654	1157	1700	10:05	31
Nov 16	0715	1158	1641	9:26	33
Dec 1	0735	1203	1630	8:56	34
Dec 16	0749	1209	1629	8:41	35

Note: Eastern Standard Time in hours and minutes on 24-hour clock. Add one hour during Daylight Time. Algonquin Park is Lat. 45°N Long. 78°W.

entire region is suitable for outdoor recreation, but precautions must be taken in winter against the likelihood of some extremely low temperatures.

Summer

It summer is defined as the frost-free season, summer lasts about four months south of the forty-fifth parallel and about three months north of this line. Summer begins about mid-May and

Table 8-2. Solar data for Boundary Waters Canoe Area, Minnesota

Date	Sunrise	Solar noon	Sunset	Day length hr:min	Twilight min
Jan 1	0757	1210	1623	8:26	36
Jan 16	0751	1216	1641	8:50	35
Feb 1	0735	1220	1705	9:30	33
Feb 16	0712	1221	1729	10:17	32
Mar 1	0647	1219	1751	11:04	31
Mar 16	0617	1215	1814	11:57	31
Apr 1	0544	1210	1837	12:53	32
Apr 16	0514	1206	1858	13:44	33
May 1	0448	1204	1920	14:32	35
May 16	0426	1203	1940	15:14	38
Jun 1	0411	1204	1958	15:47	40
June 16	0406	1207	2008	16:02	41
July 1	0411	1210	2010	15:59	41
July 16	0424	1213	2001	15:37	39
Aug 1	0443	1213	1942	14:59	36
Aug 16	0503	1211	1918	14:15	34
Sept 1	0525	1206	1848	13:23	32
Sept 16	0546	1201	1817	12:31	31
Oct 1	0606	1156	1746	11:40	31
Oct 16	0628	1152	1716	10:48	31
Nov 1	0653	1150	1648	9:55	33
Nov 16	0715	1152	1628	9:13	34
Dec 1	0736	1156	1616	8:40	36
Dec 16	0751	1203	1614	8:23	36

Note: Central Standard Time in hours and minutes on 24-hour clock. Add one hour during Daylight Time. BWCA is Lat. 48°1'N Long. 91°37'W.

lasts until the end of September in the south and lasts from mid-June to mid-September in the north and west. The shortest frost-free season is found in upland areas away from the lakes. In the Menominee Range in northern Michigan and Wisconsin, the season is less than eighty days; near Lake Nipigon north of Lake Superior, the length is down to sixty days. Summer in the Great Lakes region is characterized by warm, sunny days with frequent

Figure 8-2.   Bioclimatic index. Houghton Lake, Michigan

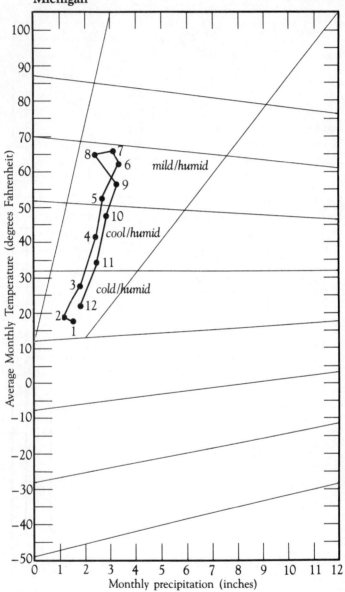

Figure 8-3.   Temperature. Boundary Waters Canoe Area

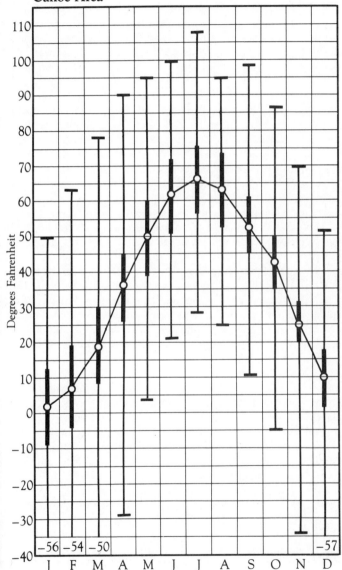

afternoon thundershowers and clear, cool nights. Afternoon temperatures average in the seventies north of the Great Lakes and everywhere close to the shore and in the eighties south of the lakes. At Houghton Lake in the interior of Michigan's Lower Peninsula, summer maximum temperatures lie between 70 and 79 degrees 44 percent of the time and between 80 and 89 degrees 41 percent of the time. They are below 70 degrees only one day in ten.

Inland from the lake shores, where continentality is marked, day-night ranges are great. Two-thirds of the days have diurnal ranges in excess of 20 degrees. Thus summer minimum temperatures are rather cool. In three-quarter of the nights, the temperature falls below 60°F; a third of the time, it falls below 50°F. The average nocturnal minimum is near 60°F in the southern portion of the region and in the low fifties in the northern and western portions.

Oppressively high temperatures are rare. Only about one day in twenty will the temperature reach above 90°F. Freezing temperatures are even more uncommon, occurring only about one day per summer.

Daytime sky cover averages about five-tenths over the region, but about two-thirds of the days are clear or partly cloudy. Daytime skies are less cloudy over and near the lakes because of the stabilizing influence of the relatively cold water. Summer cloudiness also increases somewhat along the northern and western reaches: about half the days are cloudy at Armstrong in the extreme north end of the region.

About one-third of summer days in southern portions will have some kind of precipitation; in the north, about half the days are rainy. Most summer precipitation comes in the form of showers and thundershowers. Thundershowers are more frequent in the south than in the north; at Houghton Lake, for example, about seven days per month will have thunder; at Armstrong, the frequency is nearer five per month. They are spread uniformly through the summer months.

The lakes affect the occurrence of thunderstorms. As with cloudiness, showers are less frequent on the eastern and southern shores than farther inland. Thunderstorms can be heavy, especially in the southern portions of the region when moist air from the Gulf of Mexico pushes far north. This northward flow of moist air occasionally spawns tornadoes. The region from Minnesota to Michigan lies at the northern end of "tornado alley" and has about fifty tornadoes per year. Most of these occur in June,

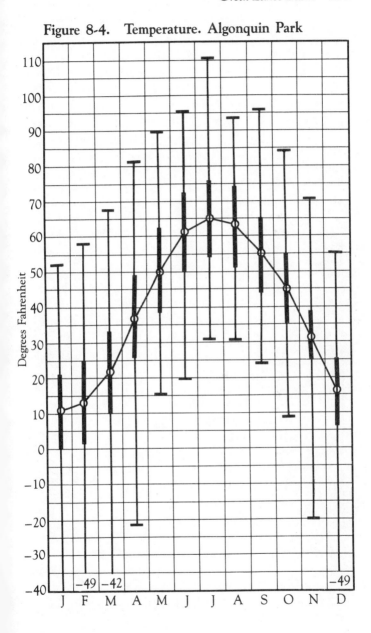

Figure 8-4. Temperature. Algonquin Park

Figure 8-5.   Sky and weather. Houghton Lake, Michigan

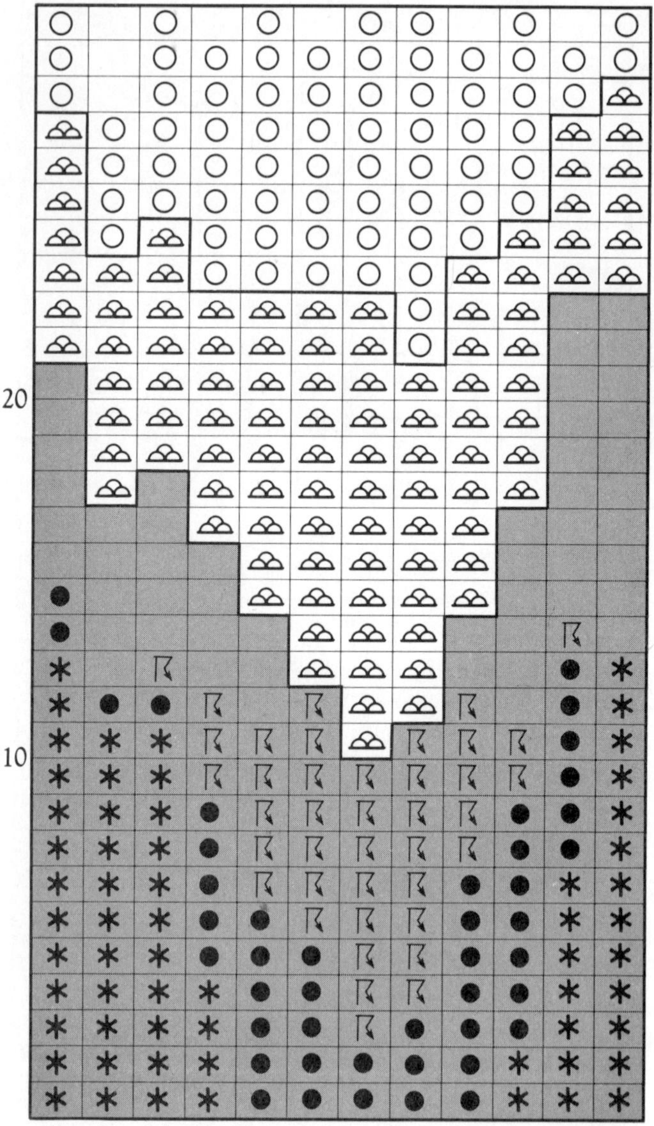

Figure 8-5. (*continued*) **Armstrong, Ontario**

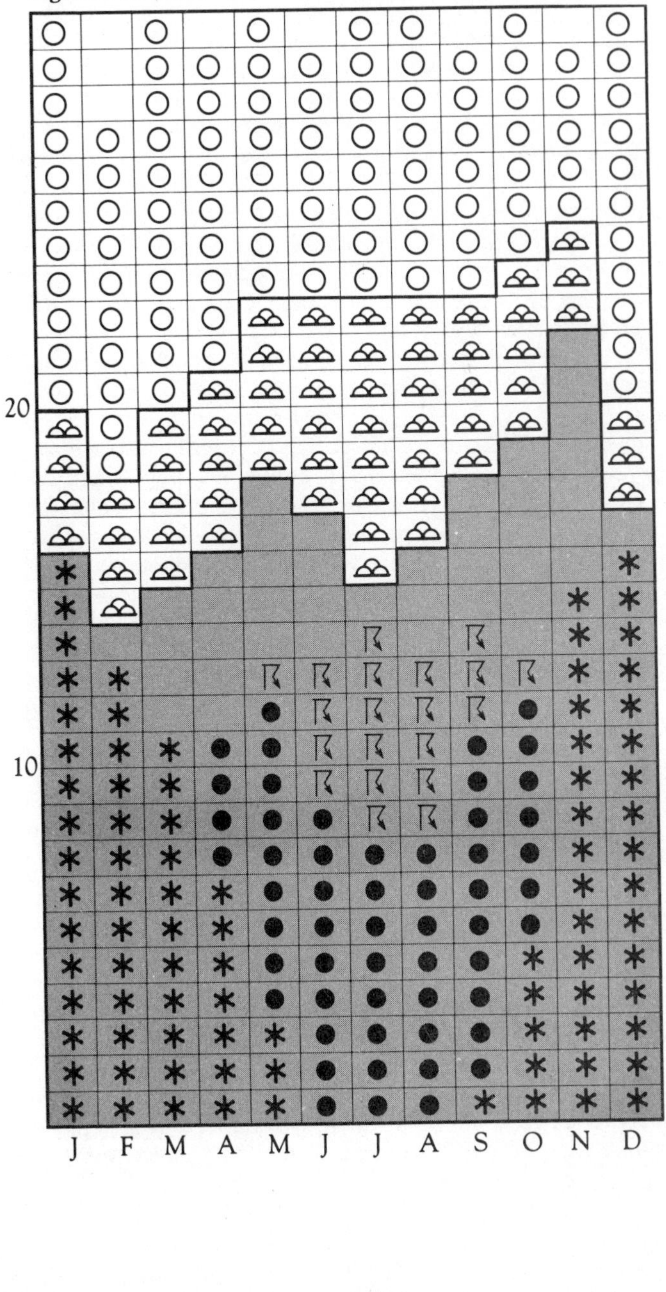

J F M A M J J A S O N D

although they can and do occur every month of the year except December. The usual season is from April through September; the usual time is between 4 and 8 P.M. The frequency decreases rapidly northward; only along the southern boundary of the Great Lakes basin are they of concern. Even there, of course, the likelihood of being caught by one is very slight.

Because most summertime precipitation is in the form of showers, the probability of several consecutive days with rainfall is slight. On the average, summer months will have one two-day rainy period. Only about one month in five will have measurable precipitation on three or more consecutive days. Long periods of dry weather are likely to prevail. Each summer month is likely to have about two periods of five or more consecutive days during which no precipitation occurs.

Fog is more frequent in the Great Lakes basin than might be expected. Advection fog, formed over the lakes when warm, moist air is transported over the cooler water surface affects shore areas downwind of the lakes, occurs during late spring and early summer. Radiation fog, caused by radiational cooling of land areas during clear nights, often appears before daybreak. In summer, fog averages about three days per month, but some locations, such as Thunder Bay and Sault Sainte Marie, may have about six days.

South winds prevail in the summer months as a result of a lobe of the Bermuda High extending inland over the southeastern United States. These winds bring in humid air from the ocean to the southern portion of the region. North and west of the lakes, summer winds are more likely to be from the southwest and west. Near the shores of the lakes, these directions are modified by daytime onshore lake breezes when the general circulation is light, about half the time during the summer. These lake breezes may penetrate 10 to 20 miles inland before dying out and are strongest in the afternoon. In open locations and over the water, these winds average about 8 to 10 miles per hour but are much less under the forest canopy. Thunderstorms may produce locally high winds and choppy waters even on small lakes. Canoists should be wary of approaching summer showers.

The summer climate of the Great Lakes region is very nearly ideal for outdoor recreation. Showers, though frequent, are of short duration. If there is a typical day, it is mild with clear or partly cloudy skies, perhaps with an afternoon shower. Nights are typically clear and cool. Adequate rain protection is necessary, but nights are generally rainless. Minimum temperatures rarely

Figure 8-6. Precipitation frequency. Atikokan, Ontario

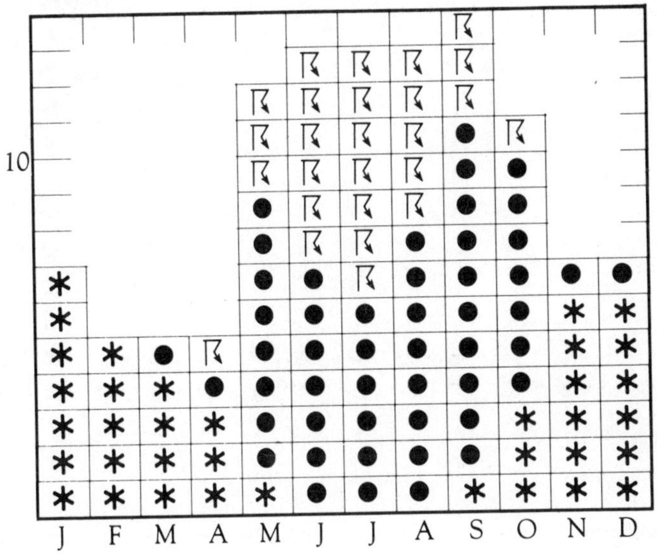

Algonquin Park, Ontario

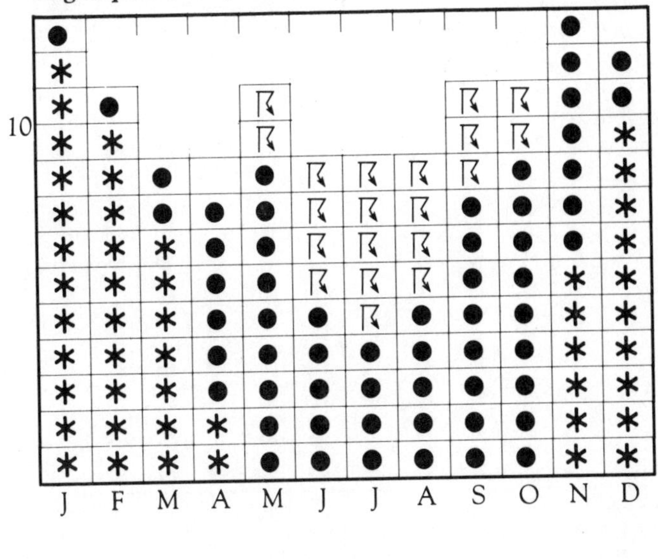

Figure 8-6. (continued) **Niagara Falls, Ontario**

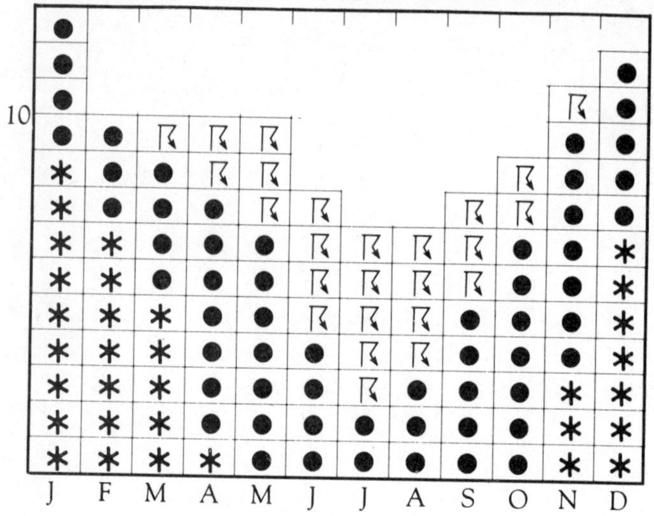

Houghton, Michigan

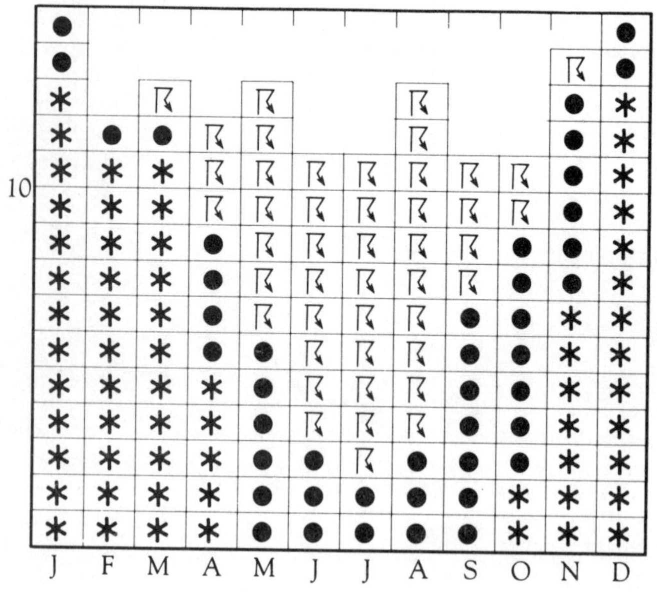

reach freezing except in low areas subject to air drainage. The Great Lakes basin offers great summer weather for backpacking and canoeing.

Autumn

As storm tracks move southward from their normal summertime position north of the Great Lakes basin, the weather becomes more variable. Frontal systems move rapidly and frequently through the area, bringing widespread showers.

Temperatures drop rapidly, until by October minimum temperatures are near 50°F in the southern portion and in the low forties in the northern portion. Minimum temperatures are in the thirties inland from the lakes and are below freezing in northern and western portions. Afternoon temperatures are on the cool side, generally near 60°F and near 50°F or below north of the lakes. The diurnal temperature range is still considerable, about 20 degrees inland, although not so great as during the summer.

Cloudiness increases rapidly as the fall wears on, blanketing the region with overcast skies on half the days by October and about two-thirds of the days by November. Nevertheless, the number of days with precipitation remains about the same. Fewer thunderstorms occur, dropping from about four per month in September to one or two in October. As colder air is drawn from the north, fall storms are increasingly likely to produce snow rather than rain. By mid-November, there is a 50 percent probability that the first snow cover of 1 inch or more will blanket the area. Even in the extreme northwestern part of the region, the first 1-inch snow cover is not likely before the first of November. But by the end of the month, nearly all precipitation comes as snow, although in the south a few rainstorms will still occur. Autumn has come to an end.

Winter

Winter takes over rapidly in December and locks the region in a deep-freeze until late March or April. November snowfall averages about 10 inches, with perhaps double that amount in the lee of the lakes. This is the start of a winter in which the average duration of snow cover ranges from about one hundred days south of the forty-fifth parallel to one hundred fifty days or more north and west of the lakes.

With the advent of snow cover, temperatures plummet. Average temperatures in January range from the twenties in Michi-

Figure 8-7. Monthly precipitation

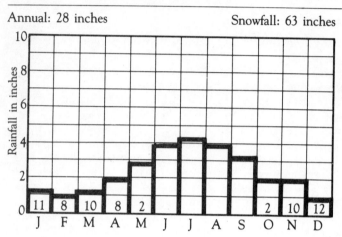

Boundary Waters Canoe Area

Annual: 28 inches   Snowfall: 63 inches

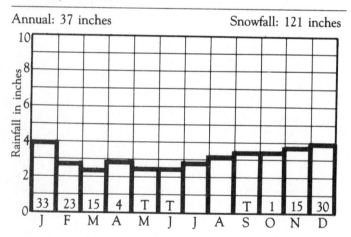

Wiarton, Ontario

Annual: 37 inches   Snowfall: 121 inches

Figures at the base of the columns indicate inches of snowfall.

Figure 8-7. (*continued*)

Algonquin Park

Annual: 34 inches   Snowfall: 99inches

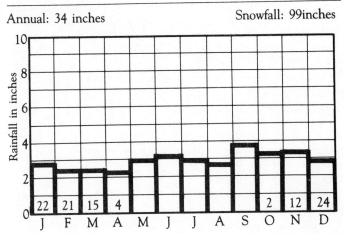

Niagara Falls

Annual: 34 inches   Snowfall: 52 inches

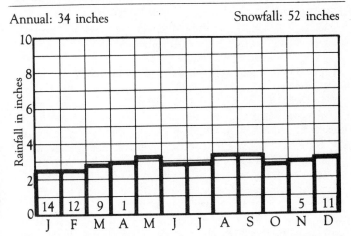

Figures at the base of the columns indicate inches of snowfall

gan and southern Ontario to 0°F or below around Lake Nipigon. In this coldest month, daily high temperatures barely reach 32°F south of the lower lakes and stay in the teens north of the lakes. Minimum temperatures are extremely low, averaging in the teens south of the lakes to 0°F and below north of the forty-fifth parallel. The coldest spot is the area north of Lake Nipigon, where the average minimum temperature is −15°F. But even this seems warm compared to the extremes of record, which are in the minus fifties. Throughout the region, extreme minimum temperatures are in the twenty- to thirty-below range.

Clouds prevail during the wintertime. Two-thirds of winter days are predominantly overcast and there are only a few clear days each month. An exception to this is the extreme north portion, where a third of the days are sunny. This area is north of the usual winter storm track and thus remains in dry continental air much of the time.

Winter snowfall amounts are substantial, ranging from 40 inches in central Michigan to 80 inches or more in the rest of the area. Much larger amounts are found downwind of the lakes in well-defined snow belts. East of Lake Ontario the winter total is more than 140 inches; similar amounts fall south and east of Lake Superior and east of Lake Huron.

In the Upper Peninsula of Michigan, there is a 90-percent probability that the snow depth will be at least 3 inches by December 1. In the northern part of the Lower Peninsula, the 90-percent probability date is mid-December. 90-percent of all winters have seen northern Michigan with a snow cover at least 1 foot deep. Cross-country skiing and snowmobiling can be counted on.

Snowfalls are frequent during Great Lakes winters. About half the days have measurable precipitation. In December and January, there is likely to be one period each month with five or more consecutive days of precipitation. But similar periods without precipitation are also likely. There will probably be one period each month of five or more days without snow. The storms tend to come in bunches.

Winds generally have a westerly component in the winter. In the middle and upper lakes region, west and northwest winds blow about half the time. Speeds are moderate, with the January average for west and northwest winds about 13–14 mph in exposed locations. Higher winds occur, of course, in winter storms and over the open water. Nevertheless, the highest winds to be expected in a typical winter will be on the order of 40–50 mph.

Figure 8-8. Mean annual snowfall in the Great Lakes area (in inches)

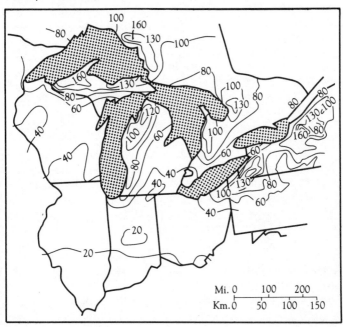

(Redrawn from Eichenlaub, Val. *Weather and Climate of the Great Lakes Region.* Notre Dame: University of Notre Dame Press. 1979)

The combination of moderate but reliable winter snowfall, low temperatures, and moderate terrain makes the Great Lakes region a paradise for cross-country skiers and snowmobilers. There are a number of downhill ski areas, but the terrain is gentle and runs are short. This is Nordic skiing country.

Spring

By April 1, snow cover in the band from Wisconsin to southern Ontario has usually disappeared. The snow cover lingers longer in the northern areas, but even there it is usually gone by May 1. Storm tracks migrate northward, although their influence is still felt in the region. Weather changes are rapid.

Temperatures moderate, and by April the mean temperature has reached into the forties except north of the forty-fifth parallel. Maximum temperatures are in the fifties. By May, temperatures are warm enough to stimulate the growth of insects; black flies and mosquitoes can be a problem. Minimum temperatures are in the thirties and below-freezing temperatures are still common.

There may be an occasional show shower in May; but by this time, nearly all precipitation is in the form of rain, even in the far northern and western areas. With the invasion of moist air from the south, showers and thundershowers become more common. In April, one or two thunderstorms can be expected; by May, the number is up to four or five. They are more frequent in the south than in the north.

The number of clear days increases, reaching a relatively stable spring and summer value throughout the region, about a quarter of the days. Overcast days decline steadily from their wintertime maximum in the south but actually increase slightly in the north as moister air replaces the predominating dry air masses of winter.

Streams are bank-full with runoff from snowmelt (and many are still used to float the winter's cut of timber to the mills downstream). Travel may be difficult in the still swampy woods, but lakes and streams are open and suitable for canoe travel. Insects are the main problem. Stock up with bug dope!

Summary

The best months for backpacking in the Great Lakes region are August and September. Insects are generally less bothersome in late summer and early fall, and the woods are somewhat drier. October is also an excellent month, especially in the southern portion. July is good, too, but may be rather heavily endowed with insects. June may find insects bad and trails wet.

Canoe routes are in their prime June through August. White-water enthusiasts may find better going in May. Fishing is said to be best from mid-May through June.

The winter skiing season normally lasts from the end of November or, more reliably, from mid-December until mid-March. There is a paucity of clear and partly cloudy days in the winter. Snow conditions are generally superb, with frequent light snowfalls and low temperatures.

Additional information

Two readily available publications contain a wealth of climatic information for the Great Lakes region. The first is more general and anecdotal; the second rather more technical with numerous climatic charts:

Eichenlaub, Val. *Weather and Climate of the Great Lakes Region.* Notre Dame: The University of Notre Dame Press, 1979.

Phillips, D. W., and McCulloch, J. A. W. *The Climate of the Great Lakes Basin.* Toronto: Environment Canada, Atmospheric Environment Service, Climatological Studies No. 20. 1972.

Although prepared primarily for tourism planning, the three volumes cited below contain a tremendous amount of summarized and analyzed climatic information:

Crowe, R. B., McKay, G. A., and Baker, W. M. *The Tourist and Outdoor Recreation Climate of Ontario.* Vol. 1: *Objectives and Definitions of Seasons;* Vol. 2: *The Summer Season;* Vol. 3: *The Winter Season.* Toronto: Environment Canada, Atmospheric Environment Service, Publications in Applied Meteorology REC-1-7s. 1977. (Available from Printing and Publishing Supply and Services Canada, Hull, Quebec K1A 0S9.)

The Michigan Weather Service has collated, analyzed, and published numerous summaries of Michigan weather. They are available from: Michigan Weather Service, 1407 S. Harrison Road, East Lansing, MI 48823. The most useful ones are:

The Climate of Michigan by Stations (with Supplements A-E); *Michigan Snow Depths;* and *Michigan Snowfall Statistics: First 1-, 3-, 6-, 12-inch Depths.*

The U.S. Environmental Data Service (National Climatic Center, Federal Building, Asheville, NC 28801) has published several useful reports in its series, *Climatic Summaries of Resort Areas, Climatology of the United States, No. 21.* The following are available:

Isle Royale National Park, Michigan. No. 21-20-1.
Houghton-Higgins Lake Recreational Area, Michigan. No. 21-20-2.
Pictured Rocks National Lakeshore. No. 21-20-3.
Sleeping Bear Dunes National Lakeshore. No. 21-30-4.

Figure 9-1. Northern Rocky Mountains

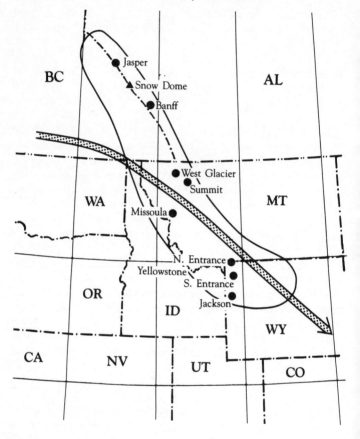

Chapter 9

Northern Rocky Mountains

As Rocky Mountain peaks go, Triple Divide Peak in the heart of Glacier National Park is not impressive. It is a spur on the ridge that stretches southeastward from Norris Mountain and stands scarcely 8,000 feet high. But from its rocky summit, one can gaze into three separate drainages that lead to the farthest shores of the continent. From the steep north face, spring snowmelt cascades into Hudson Bay Creek and flows into Saint Mary Lake before crossing into Canada. Joining the South Saskatchewan River, it finally reaches Hudson Bay at York Factory. On the western flank, Pacific Creek drains southwestward into the Middle Fork of the Flathead. After joining the Columbia near the Canadian border, the waters split the Cascades and enter the Pacific at Portland. To the east, the falling waters form Atlantic Creek, finally reaching the Missouri River and eventually flowing into the Gulf of Mexico. Triple Divide Peak is truly the apex of the North American continent.

The northern Rockies were formed by great crushing and folding forces that pushed up the ancient sediments laid down in a great inland sea. Eons of erosion have exposed the tilted layers that give the northern Rockies their characteristic appearance. The final shaping force was the action of the glaciers that scoured out the valley bottoms, giving them their typical "U" shape.

Farther to the south lie the small but magnificent Teton Mountains, which some say are the most beautiful mountains on the continent. Rising abruptly from Jackson Hole more than 7,000 feet to the summit of Grand Teton at 13,766 feet, they pose unsurpassed challenges to the mountaineer. Geologically, they are fault-block mountains, great blocks of rock thrust upward. The rock is hard, ideal for rock climbing.

Between the Tetons and the layered mountains to the north lies one of nature's greatest curiosities: the mud pots and geysers of Yellowstone National Park. The park occupies a high plateau at an elevation of about 8,000 feet above sea level. Here the fires of earth are close to the surface and fuel the fantastic array of geysers and hot springs that are the hallmark of the park.

The varied topography and the great latitudinal extent of the northern Rockies make for a varied flora. At lower elevations, ponderosa pine forms extensive open forests of stately trees. Douglas fir, Engelmann spruce, and western white pine are common at middle elevations. In some areas, extensive stands of lodgepole pine occur; it is the dominant tree in Yellowstone and accounts for about 80 percent of the forested area of the park. Near the timberline, whitebark pine, limber pine, and alpine larch are common. The forests are predominantly coniferous; but quaking aspen, balsam poplar, and paper birch add variety, especially with their bright fall colors.

The timberline varies markedly with latitude. In the far north, the upper limit of tree growth is little more than 5,000 feet above sea level. In Glacier Park, timberline lies at about 7,000 feet; and in the Tetons, it is close to 10,000 feet.

General climate

The dominant features of the northern Rockies climate result from the interaction between the topography and the main flow of air from the west. Although the coastal mountains take a large part of the moisture from the Pacific air masses as they move eastward, there is still a substantial amount left. Immediately to the east of the Cascades, there is a pronounced rain shadow where the annual precipitation is rather low—around 10 inches. The cause of this is not so much that the mountains have removed much of the water from the air mass but rather that the downslope motion of the air raises the air temperature, lowers its relative humidity, and generally inhibits the upward motions necessary to produce rain.

Sierra Club membership supports vital programs on behalf of wilderness conservation and a more healthy environment.

Sierra Club

MEMBERSHIP FORM

☐ *Yes, I'm proud to join!*

New Member Name _____

Address _____

City/State _____ Zip _____

☐ **GIFT MEMBERSHIP:** *Sign up my friend!*
Here is my name for gift card acknowledgement.

_____

Donor Name

_____

Address   City/State   Zip

MEMBERSHIP CATEGORIES

	Individual	Joint
REGULAR	☐ $ 33	☐ $ 41
SUPPORTING	☐ $ 50	☐ $ 58
CONTRIBUTING	☐ $100	☐ $108
LIFE	☐ $750	☐ $1000
SENIOR	☐ $ 15	☐ $ 23
STUDENT	☐ $ 15	☐ $ 23
LIMITED INCOME	☐ $ 15	☐ $ 23

All dues include subscription to *Sierra* ($7.50) and chapter publications ($1).
Dues are not tax deductible. **Enclose check and mail to:**

Sierra Club

J-915, P.O. Box 7959
San Francisco, CA 94120

There is still moisture left in the air that has traveled over the far western mountains when it passes over the Rockies. In the Tetons and Glacier Park, the annual precipitation may be as high as 60 inches. Farther north, along the British Columbia-Alberta border, annual totals are generally somewhat less, around 50 inches. But there is a great deal of variation, even in relatively small geographic areas. A detailed analysis of the precipitation in Yellowstone National Park, only about 60 miles square, indicated a maximum in excess of 80 inches on the windward side of the higher elevations to less than 20 inches per year in the dry valleys in the northern part of the park. Generalizations about precipitation amounts are risky.

Nevertheless, some useful generalizations can be made. The main winter storm track crosses Montana on a northwest to southeast course. Areas south of the low centers will produce their precipitation in winds from the southwest. North of the center, winter storm winds are more likely to have an easterly component. Thus on the eastern side of the range, precipitation amounts are somewhat higher in the north than in the south, about 20 inches compared with 15 inches. The highest precipitation amounts are always found on the western slopes. Just west of Banff and Jasper parks, in the Selkirk Mountains, annual precipitation amounts greater than 50 inches are found in the valleys; the amounts at higher elevations are likely to be considerably greater. This precipitation, mostly in the form of snow at the high elevations, feeds the great Columbia Icefield.

Farther south, at Waterton-Glacier International Peace Park, precipitation amounts are in excess of 50 inches. In the Bitterroot Range, along the Idaho-Montana border, precipitation amounts range to 60 inches. Yellowstone Park is relatively dry, with the park average near 30 inches; but the Tetons to the south again show a maximum in excess of 50 inches. The Wind River Mountains, southwest of the Tetons, lie partly in the rain shadow and generally have amounts less than 40 inches. The Bighorn Mountains to the east are shortchanged and can barely squeeze 30 inches from what is left.

In the lowland areas in the rain shadow of all these ranges, rainfall amounts drop to near desert levels. Between the Wind River and Big Horn Mountains, for example, there is a large area in which the annual precipitation is less than 8 inches. Even farther north, the annual amounts in the lee of the mountains are generally less than 20 inches.

Figure 9-2. Monthly precipitation

Jasper, Alberta

Annual: 16 inches   Snowfall: 54 inches

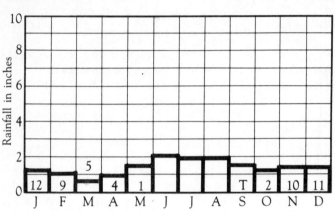

Yellowstone National Park (north entrance)

Annual: 17 inches   Snowfall: 90 inches

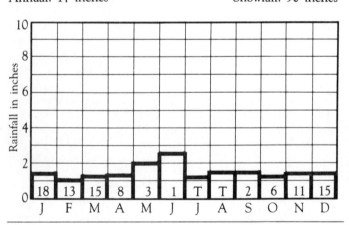

Figures at the base of the columns indicate inches of snowfall.

Figure 9-2. (continued)

Summit, Montana

Annual: 37 inches   Snowfall: 256 inches

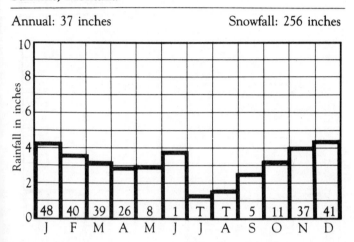

	J	F	M	A	M	J	J	A	S	O	N	D
	48	40	39	26	8	1	T	T	5	11	37	41

Yellowstone National Park (south entrance)

Annual: 33 inches   Snowfall: 217 inches

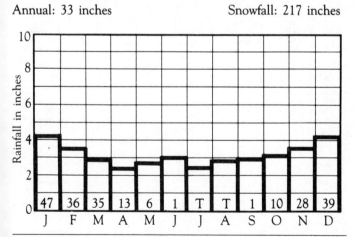

	J	F	M	A	M	J	J	A	S	O	N	D
	47	36	35	13	6	1	T	T	1	10	28	39

Figures at the base of the columns indicate inches of snowfall.

Figure 9-3.   Temperature. Banff, Alberta

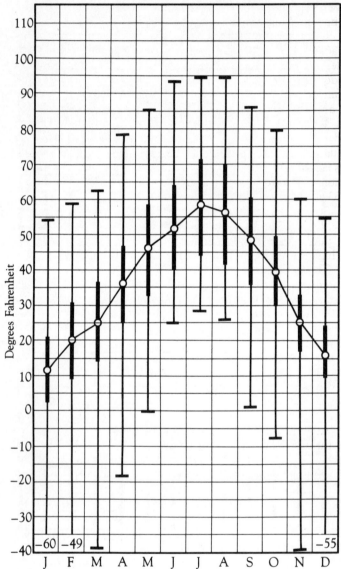

The annual cycle of precipitation is rather even: most areas have a summer minimum barely below the annual mean. In the far north, there is a slight summer maximum.

The entire region displays the large diurnal and annual temperature ranges characteristic of continental climates. The range is less in the wetter regions than in the drier because of the moderating effect of cloud cover. The daily range is usually more a function of local microclimate than of latitude or location vis-à-vis the crest. The lowest temperature recorded in the northern Rockies was a chilling −70°F, at Rogers Pass in the southern end of the Lewis Range not far from Helena, Montana, on January 20, 1954. On February 9, 1933, a low reading of −66°F was recorded in Yellowstone Park. Yet in the Canadian end of the region, the lowest temperature of record was −63°F in January 1953 at Lake Louise.

Summer maximum temperatures are somewhat more bearable. Daily maxima are generally in the seventies: low seventies in the north, high seventies in the south, during July and August. At high elevations, the temperatures are somewhat lower.

The combination of substantial rainfall amounts and low winter temperatures means glaciers are abundant, especially from Glacier Park north. The level of perpetual snow is about 8,000 to 9,000 feet in Glacier National Park and about 1,000 feet lower in Jasper National Park. This does not mean above this level all is snow; it means snow will persist year-round in favorable locations.

Cloudiness is rather high and uniform throughout the year in the northern end of the region. The farther south one goes, the greater the tendency toward sunny summers and cloudy winters.

Wind speeds are moderate and at lower elevations are controlled primarily by topography. Mountain and valley circulations are the rule in fair weather: generally upvalley in the daytime and downvalley at night. Above the timberline, high winds may occur, especially in stormy weather; but the Rockies are not prone to exceptionally strong winds. A Canadian climatological analysis indicates the highest gusts in the northern Rockies may be about 80 mph.

In late winter and early spring, exceptionally dry winds may blow eastward down the mountain slopes. These "chinook" winds are referred to locally as "snow eaters." They are so hot and dry they can make a foot of snow disappear in a few hours. They are caused by air that has come from the Pacific and been dried out by

Figure 9-4.   Temperature. Yellowstone National Park

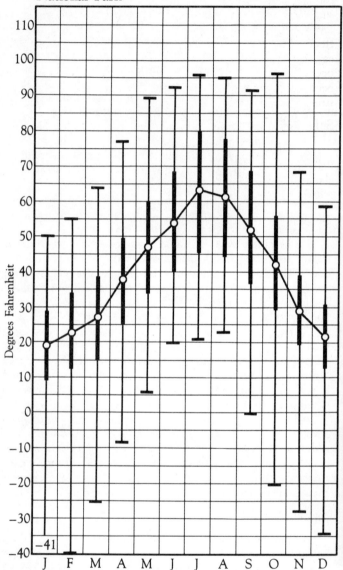

passages over three mountain ranges: the Coast, Cascade, and Rocky mountains. The condensed water has also added heat to the air, which is further heated by adiabatic compression as it slides down the eastern slope; it arrives dry and warm at low levels. The boundary between the warm, dry air sliding down the Rockies and the relatively cold air occupying the plains to the east may be very unstable. Sharp temperature variations may occur within a very few minutes. In Alberta, the temperature has been known to rise 60 degrees in one hour. An exceptionally strong chinook was the culprit.

Solar data

Solar data for West Glacier, Montana are presented in Table 9-1. Winter day length is about 8½ hours; summer day length is about 16 hours. At Jasper winter days are shorter by about 47 minutes; summer days are longer, about 54 minutes.

Bioclimatic index

Compared with the coastal mountains to the west, the Rockies are much colder and drier. At Marias Pass, Montana (elevation 5,213 feet), winter months qualify as cold/wet—zone I on the bioclimatic chart. June, a rather wet month, is classed mild/humid; July and August are substantially drier—mild/dry. Transition months are cool/humid.

Farther south, at Yellowstone, precipitation amounts are less and the bioclimate ranges from cold/humid in the winter to mild/dry in the summer. At Banff, the climogram is similar to the Yellowstone climate although about 7 degrees colder. In figure 9.5 it would have the same shape as the one shown for Yellowstone, but displaced downward about 7 degrees.

Above the timberline, the climograms will be displaced downward about 3 degrees for each thousand feet of elevation. In addition, stronger winds will cause the windchill temperature to be substantially lower, dropping winter bioclimate into the very cold and bitterly cold zones. Winter above timberline requires full arctic equipment.

Summer

"Warm in the sun and cool in the shade" characterizes the half or more summer days that are predominantly sunny. Oppressively high temperatures are nonexistent. Record high temperatures at Jackson, Wyoming, Marias Pass, Montana, and Banff,

Table 9-1.   Solar data for West Glacier, Montana

Date	Sunrise	Solar noon	Sunset	Day length hr:min	Twilight min
Jan 1	0828	1239	1651	8:23	36
Jan 16	0822	1246	1709	8:46	35
Feb 1	0806	1250	1733	9:28	33
Feb 16	0742	1250	1758	10:15	32
Mar 1	0716	1248	1820	11:04	31
Mar 16	0646	1244	1843	11:57	31
Apr 1	0613	1240	1907	12:54	32
Apr 16	0543	1236	1929	13:46	33
May 1	0515	1233	1951	14:35	35
May 16	0453	1232	2011	15:18	38
June 1	0438	1234	2030	15:52	40
June 16	0433	1237	2040	16:07	41
July 1	0438	1240	2041	16:03	41
July 16	0452	1242	2032	15:41	39
Aug 1	0511	1242	2013	15:02	36
Aug 16	0532	1240	1948	14:17	34
Sept 1	0554	1236	1917	13:24	32
Sept 16	0615	1231	1846	12:31	31
Oct 1	0636	1225	1815	11:39	31
Oct 16	0658	1221	1745	10:46	31
Nov 1	0723	1220	1716	9:53	33
Nov 16	0746	1221	1655	9:09	34
Dec 1	0807	1225	1643	8:36	36
Dec 16	0822	1232	1641	8:19	36

Note: Mountain Standard Time in hours and minutes on 24-hour clock. Add one hour during Daylight Time. West Glacier, Montana is Lat. 48°30′N Long. 113°59′W.

Alberta, are all 94°F. More typical are the average maximum temperatures in July: 81°F at Jackson, 74°F at Marias Pass, and 72°F at Banff.

Because the climate is rather dry, with nights predominantly clear, strong radiational cooling produces low nighttime temperatures and unusually large diurnal temperature ranges. Diurnal ranges of 30 degrees to 35 degrees are common. Thus a day with a comfortable afternoon temperature of 70°F is likely to drop to

Figure 9-5. Bioclimatic index

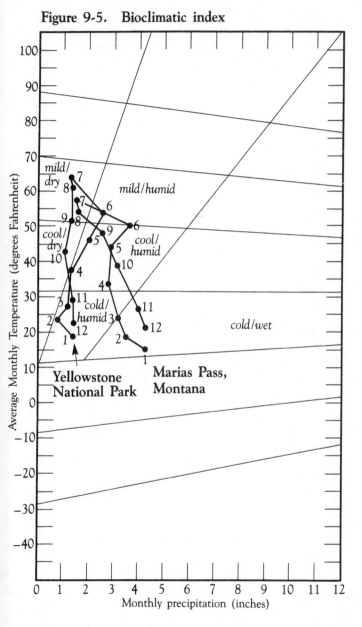

Figure 9-6. Occurrence of high and low temperatures. Yellowstone National Park

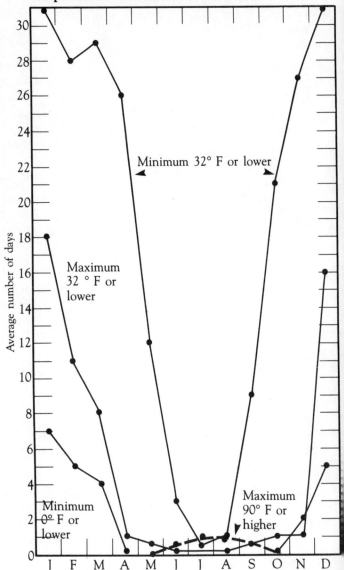

near freezing at night with clear skies. In low areas where cold air can accumulate, summertime frosts are common. Paradoxically, the air drainage that keeps low elevations cold keeps higher elevations relatively warm.

At Yellowstone, summer months average one to three nights with below-freezing temperatures (see figure 9-6). In May and September, the numbers are twelve and nine, respectively.

Few records are available for the higher elevations. At Sperry Chalet, high above Lake McDonald in Glacier National Park (elevation 7,000 feet), the maximum July and August temperatures during one summer of observations were 84°F and 75°F, respectively. At West Glacier, 4,000 feet lower, the maximums were 10 degrees warmer.

Sunny weather is common in the northern Rockies, especially in its southern end. At Missoula, Montana, half of July and August days are clear and another quarter are partly cloudy. Overcast skies occur only one day in four. Farther north, the frequency of cloudy days increases: at Jasper, half the days are cloudy and fewer than a quarter are clear. July is the sunniest month at Glacier; July and August are the sunniest months at Jasper.

Along with the sunshine come low humidities. Afternoon relative humidity at Missoula averages 30 percent and seldom goes over 40 percent. This low humidity combined with sunny weather produces frequent periods of high fire danger in the Rockies, especially in Wyoming and Montana. The great Idaho fire of 1910 that consumed more than 3 million acres of timberland started on August 10. Typically, fire danger peaks in July and August—the best hiking weather is the worst fire weather.

It should come as no surprise that precipitation amounts are low in the summer throughout the region. At Yellowstone, monthly amounts range between 1.5 and 2.5 inches. These amounts are typical of the entire region. Some summers are almost completely rain-free, contributing greatly to the fire hazard and a scarcity of drinking water on some trails. Wet summers are nonexistent. The wettest summer months on record at Missoula have had no more than 4 inches of rain; the wettest August had only 2.5 inches. Most of it comes in brief showers, which are occasionally heavy, the entire month's rainfall coming in two or three showers. Although summer months typically have an average of six thunderstorms, many of these are "dry," with no rain reaching the ground.

Figure 9-7.   Sky and weather. Jasper, Alberta

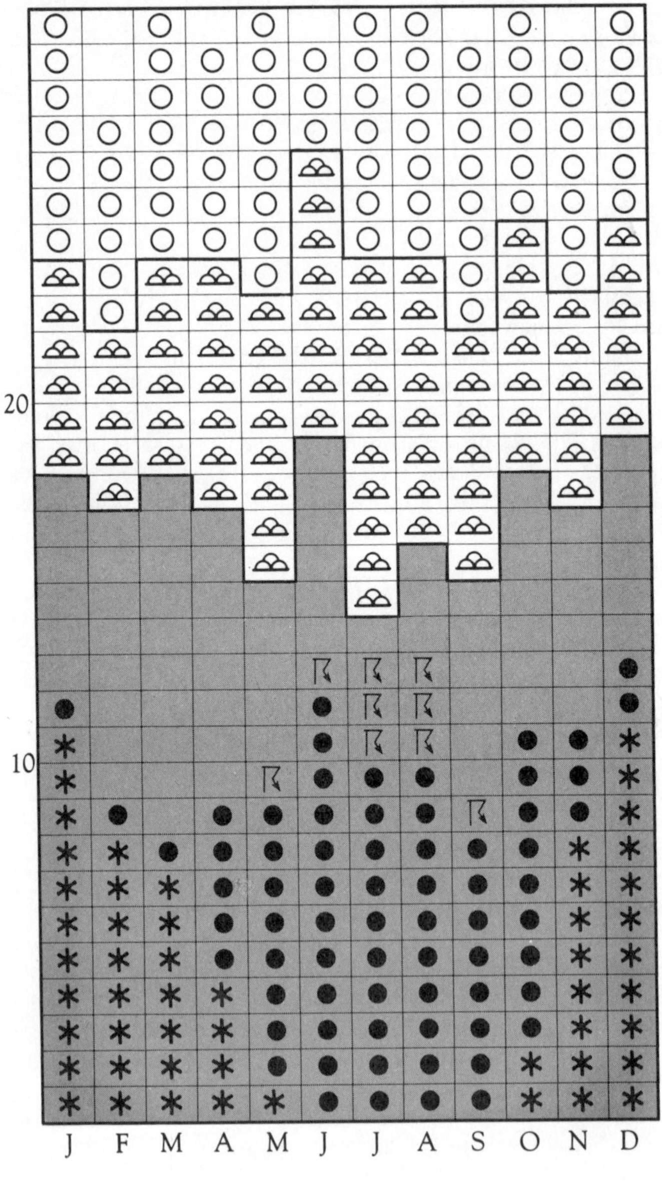

Figure 9-7. (continued) **Missoula, Montana**

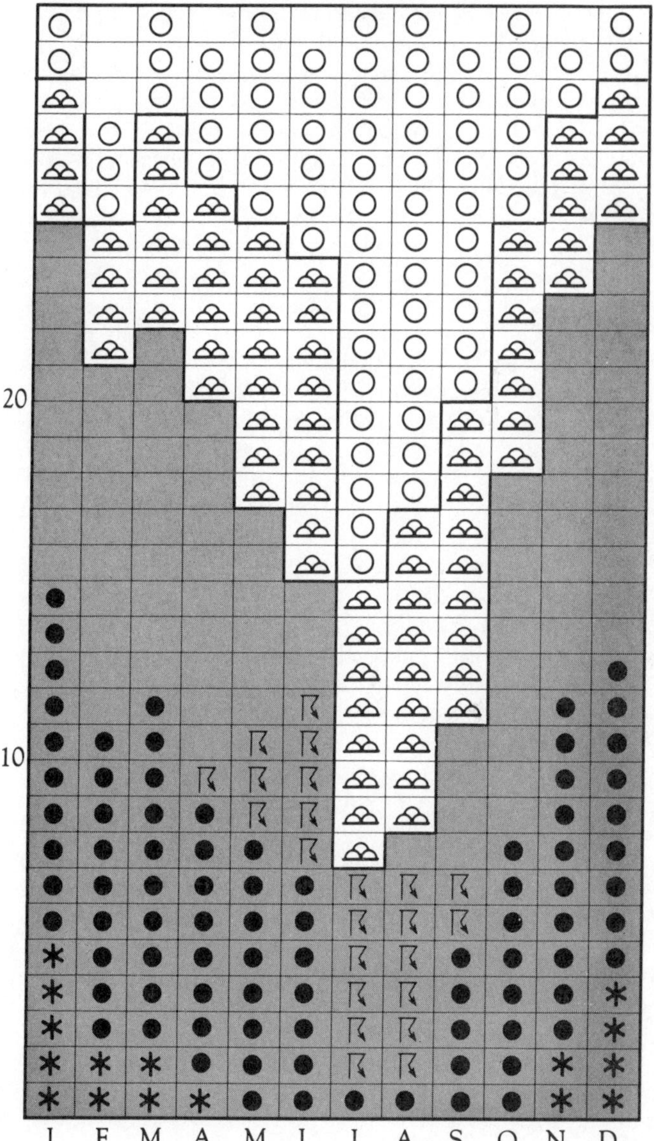

Farther north, in the Canadian Rockies, thunderstorms are less frequent but rainy days occur more often. Rainfall amounts are light: Jasper's 2 inches per month is spread over thirteen days.

One caution: snow showers can occur any time during the summer, especially at elevations above 6,000 feet. Polar air occasionally invades the region from the north, triggering snow showers that can drop several inches in a brief time. Four inches were recorded in Yellowstone in July 1913, 1921, and 1943. Banff has received a 5-inch snowfall in June and a 3-inch snowfall in August. Summer snow is not uncommon above 9,000 feet.

Most areas below 7,000 feet are free of snow by late June. At higher elevations, persistent snowpacks in the passes and on shaded slopes can create problems for backpackers. July is a fine month for hiking; but trails muddy from snowmelt may make walking somewhat messy, and high water in creeks may make bridgeless crossings hazardous. Streams have usually returned to normal by the end of July and may even start to dry up, especially those not fed by glacial melt. Trail and stream conditions should be checked locally before venturing into the high country early in the season.

Wind is rarely a problem in the summertime. Local winds predominate. The jet stream is normally far to the north, and even the high peaks have only light to moderate winds. Thunderstorms are often accompanied by brief periods of high winds that can whip up whitecaps on lakes very quickly. Boaters and canoeists should be especially alert to the approach of thunderstorms and take shelter before the storm breaks.

August and early September are thus the best months for high-mountain hiking, with July close behind except in years with heavy and long-lasting snowpacks. Above the timberline, thunderstorms and the occasional snowstorm are the major hazards; the backpacker should be alert to threatening weather signs. Towering cumulus clouds with high bases, often preceded by high puffy cirrocumulus, are harbingers of dry lightning storms, which often produce much cloud-to-ground lightning. Most of the time, however, the weather is benign and ideal for backpacking. Above the timberline, the hiker must always be alert to the dangers of hypothermia and should be equipped with suitable clothing.

Autumn

It has been said that there are only two seasons in the northern Rockies: winter and July. Although that may be something of an exaggeration, winter snows often start by late September or early October. Consider that while the Washburn expedition of 1870 was exploring Yellowstone, a 2-foot snowfall blanketed the park in mid-September. One of the members of the party became separated from the main group and wandered for thirty-seven days without adequate equipment or clothing. Fortunately, the weather turned warm, the snow melted quickly, and he was eventually able to rejoin the expedition. Such a large early-season snowfall has not been recorded since (and records have been kept since 1877), but the possibility of a similar occurrence remains. Generally, though, September and early October have good hiking weather at low elevations.

Daytime temperatures are cool, reaching the low sixties in September but dropping to the fifties by October. Nights are crisp, with freezing temperatures and readings in the teens common.

Precipitation remains light but will often start as rain and change to snow as the storm passes and starts to pull in colder air from the north. At higher elevations, snow will start to accumulate. Cloudiness increases, and by October, half the days are overcast. Fog may fill the western valleys and lakes during late autumn. Ice will begin to form on small lakes by the end of October.

Backpacking is usually possible at low elevations through October, but the threat of snow and cold weather dictate preparedness for alpine conditions. Above the timberline, winter has usually set in by November and winter mountaineering equipment and experience are mandatory.

Winter

Snow falls in substantial amounts in every month from October through May, although the maximum accumulation is usually reached by late March or early April. The snowpack is usually well established by November; many secondary roads are not plowed and are closed to traffic after November 1. Temperatures drop into the teens regularly by the end of November and to near 0°F or below by mid-December. At Yellowstone, minimums below 0°F can be expected on five or six days in December, January, and February. Minimum temperatures of −40°F occur regularly every winter, although the average minimum is near 0°F. The record low temperature for the forty-eight coterminous states is −70°F at Rogers Pass, south of Glacier Park. Only Prospect Creek in Alaska and several locations in the Yukon have experienced lower temperatures, with the lowest at Mayo in the Yukon, −81°F.

Daytime temperatures average about 20 degrees warmer than nighttime minimums. In the deep winter months, maximum temperatures average below freezing throughout the region.

The lack of winter sunshine contributes to the bitterness of the winter climate. In the north end of the region, more than half the days are overcast; only one day in four qualifies as sunny. Farther south the picture is even gloomier: at Missoula, twenty-five days out of thirty-one in December and January are predominantly cloudy and only two days per month are clear.

With annual precipitation amounts rather modest, ranging from 20 to 40 inches and spread nearly uniformly throughout the year, it might be expected that winter snowfall amounts would be low. Such is not the case, however, for nearly all winter precipitation falls as low-density snow. At Marias Pass in Glacier, about 25 of the 37 inches of annual precipitation fall as snow, producing 256 inches total snowfall. (Each inch of melted precipitation produces about 10 or 11 inches of snow.) Although records from high elevations are not available, it is likely that many above-timberline areas, especially on the western slopes, receive more than 1,000 inches of snow in a typical winter.

Thus most mountain slopes accumulate depths of 8 to 10 feet or more by the beginning of the melt season in March or April. At lower elevations, the maximum snow depth will be somewhat less. At Marias Pass (elevation 5,213 feet), the maximum depth of snow on the ground during twenty-six years of observation ranged from 26 to 91 inches, with an average maximum depth of 55 inches. At West Glacier (elevation 3,154 feet), the average maximum depth for the same period was 25 inches and the range was 7 to 41 inches.

Greatest snowfall totals are found west of the continental divide. The rain shadow (more properly, snow shadow) effect is pronounced. The Selkirks west of the divide receive more than 120 inches of snowfall per winter; but Jasper and Banff, east of the divide, receive an average of 54 inches and 86 inches, respectively. Figures on average snowfall amounts tend to be misleading, however. The interaction of local topography and direction of winter storm winds can produce remarkable variations in short distances. Yellowstone Park snowfall has annual amounts ranging from 50 inches at Gardner on the northern boundary to 600 inches near the summit of 10,250-foot Mount Sheridan.

At low elevations, winter precipitation can be either snow or rain. Above the timberline and even at low elevations in the Canadian Rockies, most of the precipitation is in the form of snow from November through March. Cold waves accompanied by subzero temperatures and strong winds with blowing snow occur in the northern portion of the region about six to twelve times a year and are especially evident east of the continental divide. Between these cold waves, temperatures moderate; and there is often a period of a week or ten days of relatively mild and windy weather. This is the so-called chinook weather, and it can bring winds of 25 to 50 mph that blow for several days at a time. The chinook is largely confined to the slopes east of the divide and will typically be associated with snow or rain west of the divide.

The northern Rocky Mountain high country in the winter is the province of the ski mountaineer. Avalanches are common and occasionally fatal. The worst mountaineering avalanche accident occurred on the slopes of Mount Cleveland in the north end of Glacier Park on December 29, 1969. Five young climbers attempted to scale the peak against the advice of a park ranger. A massive avalanche on the west face caught all five and buried them so deep that their bodies were not recovered until the following July. Nevertheless, with proper conditions and precautions, winter mountaineering in the northern Rockies can be challenging and relatively safe.

Spring

Spring comes late to the northern Rockies. The snowpack has generally disappeared from the low elevations by early May but remains in some heavily drifted areas in the high country until mid-August. Spring snowmelt makes for muddy going on many trails; the trails may not dry for several weeks after the snow has gone. Many secondary roads across the passes and in the backcountry are not opened until the end of May. Streams are bank-full from snowmelt and crossings may be hazardous in late June. High-country travel may thus be difficult through June and sometimes into early July. Wet snow in May and June is common in the high country but rare at low elevations.

Spring is a time of rapid and abrupt weather changes. Thunder, lightning, rain, snow, hail, sleet, and wind blowing from 0 to 50 mph have all been recorded in one hour.

Spring is also the time of the mosquito in the Rockies. Snowmelt provides numerous pools and boggy places for the insect to breed. Mosquitoes can be very bothersome until snowmelt is completed, usually about July 1. The same conditions that produce the muddy trails and the mosquitoes also produce a profusion of wildflowers, especially above the timberline.

Spring is not a particularly propitious time for backpacking. The hiker must be prepared for mud, snow, high water, mosquitoes, and rapid weather changes. But the rewards can be great if you hit the high country at the peak of the alpine flower bloom.

Summary

The best time for hiking the high country is from mid-July to mid-September. A low snowpack from the previous winter can sometimes extend this a month or so into the late spring. In the low country, September and October can be fine months for backpacking; but the hiker must be alert and prepared for an early onset of winter snow.

Winter is long and snowy and produces excellent conditions for downhill skiing and ski mountaineering. Snowfalls are lighter and less frequent than in the Cascades and Olympics to the west. The snow is fluffier and more to the liking of the powder buffs.

Spring, which may last until early July, is the least desirable hiking season; but high-altitude skiing can be excellent, especially on the glaciers in Banff and Jasper parks.

Additional information

An excellent, comprehensive report on the recreation climate of Yellowstone National Park and Grand Teton National Park is in the process of being published by the Park Service. Inquiries should be directed to Superintendent, Yellowstone National Park, Wyoming 82190. The full report (which may be available from the Department of Atmospheric Science, University of Wyoming, Laramie, WY 82071) is:

Dirks, R. A. and Martner, B. E. *The Climate of Yellowstone and Grand Teton National Parks.* U.S. Dept. of Interior, National Park Service, Natural Resources Report. August 1978.

A less comprehensive climatic guide is available from Mountain States Weather Services, 904 East Elizabeth Street, Fort Collins, CO 89521:

Wirshborn, James E. *Climate of Yellowstone Park. A Visitor's Guide to Yellowstone Seasons.* Ft. Collins: Mountain States Weather Services. 1978.

For Glacier National Park, the following is useful:

Dightman, R. A. *Climate of Glacier National Park.* Glacier Natural History Association Bulletin No. 7. 1961. (Available from the Association at West Glacier, MT 59936.)

Two Canadian publications contain information on the recreation climate of the Canadian Rockies:

Bennett, Richard C. *Climatic Suitability for Recreation in British Columbia.* Victoria: Ministry of the Environment, Resource Analysis Branch. January 1977.

Masterton, J. M., Crowe, R. B and Baker, W. M. *The Tourism and Outdoor Recreation Climate of the Prairie Provinces.* Toronto: Atmospheric Environment Service, Meteorological Applications Branch, Publications in Applied Meteorology REC-1-75. 1976.

Figure 10–1. Southern Rockies

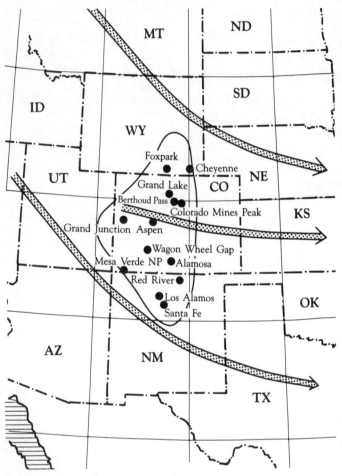

Arrows indicate winter storm tracks.

Chapter 10

Southern Rocky Mountains

Just west of the 105-degree meridian, the great Front Range of the Rocky Mountains stretches in a nearly unbroken facade for 500 miles. Starting from the level plains in eastern Wyoming, Colorado, and New Mexico, the mountains rise abruptly to the continental divide a few miles to the west. From Milner Pass in Rocky Mountain National Park to the New Mexico border, no pass across the divide is lower than 10,000 feet; most are above 11,000 feet. Twenty peaks along the divide are more than 14,000 feet high. There are almost nine hundred peaks in the southern Rockies over 11,000 feet high; fifty-four of these are more than 14,000 feet.

Most of the ranges making up the southern Rockies are oriented more or less north-south. Between the ranges lie the high parks—great meadows with occasional clumps of pines at elevations of 8,000 to 11,000 feet. The true timberline is higher, ranging from 11,000 feet in the north in the Medicine Bow Mountains in southern Wyoming to nearly 12,000 feet in the southern Sangre de Cristos in northern New Mexico.

The southern Rockies are the highest mountain mass in the coterminous United States, containing three-quarters of the land area above 10,000 feet. It is not surprising, then, that the rivers flow every which way from the center. The North Platte River flows northward from northern Colorado before bending eastward at Casper, Wyoming, to form the northern boundary of the southern Rockies. The Colorado River rises just a few miles away and flows generally westward to the Utah border, where it turns south to meet the Green River in Canyonlands National Park. The Rio Grande also has its source in the Rockies, a few miles west of Alamosa, Colorado. Many tributaries of the Mississippi flow eastward from the Front Range.

The southern Rockies cover nearly 100,000 square miles. Within this area, there are two national parks, two national monuments, eleven national forests, twenty wilderness areas (with more to come), more than sixty ski areas, and thousands of miles of hiking trails. Climate and topography conspire to provide some of the best hiking and skiing on the North American continent. Some say it is the best.

General climate

The two major controls over the climate of the southern Rockies are their location in the central part of the continent and their extreme topographic variability. The former implies a continental climate with large diurnal and seasonal variability. The latter produces large elevational and microclimatic changes. It is thus difficult to speak of "the climate" of the Rockies. The climate is a mosaic of microclimates and mesoclimates, with great variations sometimes occurring in a few hundred feet of horizontal or vertical distance.

The main features of the general circulation that control storm movements provide the setting within which the local climates develop. Summer climate is dominated by high pressure to the east and low pressure to the south and west, which floods warm, moist air from the Gulf of Mexico throughout the Great Plains. This provides a supply of moist air to feed the numerous summer thunderstorms that occur in the mountains. Eastward-moving air from the Pacific Ocean reaches the Rockies, but it is bereft of most of its moisture after its long journey over the Sierra Nevada and other mountain ranges.

In the winter, however, Pacific storms penetrate far into the continent, often with enough moisture to produce significant amounts of snowfall in the high mountains. These storms generally travel southeastward, following tracks either north or south of the Rockies (figure 10.1). Most of their residual moisture is deposited as snow on the higher elevations and on west-facing slopes. Relatively little falls on the eastern slopes.

When cold polar air pushes southward in the plains and contacts warm, moist air from the gulf, a low-pressure storm often forms on the front. When such a storm develops in the western plains, snow, rain, or both are dumped on the eastern slopes, often in substantial quantities.

Annual precipitation amounts are relatively low, about 20 inches at valley stations and about double this at high elevations. There is also a weak north-south gradient (20 inches at Grand Lake, Colorado, 18 inches at Los Alamos, New Mexico) and a stronger east-west gradient (19 inches at Aspen, Colorado, 8 inches at Grand Junction, Colorado).

The main axis of the winter jet stream frequently positions itself over the southern Rockies with the upper air flow from the northwest. Although there often is no well-defined frontal system or low-pressure area associated with this flow, considerable amounts of moisture can be squeezed out of the air flowing southeastward over the mountains. Heavy precipitation, generally in the form of snow, occurs near the continental divide. With strong west-to-east flow, Pacific Coast storms may penetrate the continent with sufficient moisture to produce substantial snowfalls west of the divide also.

When the jet stream courses southward west of the mountains and loops northward east of the mountains, a potential for very large precipitation amounts exists. A low-pressure center often forms in western Colorado and moves eastward relatively slowly. The combination of frontal activity and orographic lifting can produce storm precipitation amounts in excess of 14 inches of snow. A storm of this type on April Fool's Day in 1957 was no joke. It dumped 3 feet of snow on Berthoud Pass; the total for the first two weeks of that month was 122 inches.

Winds are high on the above-timberline peaks, averaging 26 mph for the year on top of Colorado Mines Peak (elevation 12,493 feet). There is relatively little variation from winter to summer: January's mean wind speed at this location was 34 mph; the estimated July speed was 21 mph. These compare with 46 mph and 25 mph, respectively, at the summit of Mount Washington in New Hampshire (elevation 6,288 feet).

Solar data

Data applicable to the Southern Rocky Mountain region are presented in Table 10-1, for Aspen, Colorado. Times of sunrise, solar noon and sunset vary only a few minutes throughout the area. Day length in the northern portion is about ten minutes shorter in winter, the same amount longer in summer. In the southern portion, days are about ten minutes longer in winter, and ten minutes shorter in summer.

Bioclimatic index

In the Rockies, the bioclimate is greatly dependent on elevation and exposure. At Grand Lake, at the western entrance to Rocky Mountain National Park, winter bioclimate is cold/humid (figure 10.2). A few miles away, on the summit of Colorado

Table 10-1. Solar data for Aspen, Colorado

Date	Sunrise	Solar noon	Sunset	Day length hr:min	Twilight min
Jan 1	0727	1211	1655	9:28	30
Jan 16	0725	1217	1709	9:44	30
Feb 1	0715	1221	1727	10:13	29
Feb 16	0658	1221	1745	10:46	28
Mar 1	0639	1220	1800	11:21	27
Mar 16	0616	1216	1815	11:59	27
Apr 1	0551	1211	1831	12:40	27
Apr 16	0528	1207	1846	13:18	28
May 1	0508	1204	1900	13:52	29
May 16	0453	1204	1915	14:22	31
June 1	0443	1205	1928	14:45	32
June 16	0441	1208	1936	14:55	33
July 1	0445	1211	1937	14:52	33
July 16	0455	1213	1932	14:38	32
Aug 1	0508	1214	1919	14:11	30
Aug 16	0522	1211	1901	13:39	29
Sept 1	0536	1207	1838	13:01	28
Sept 16	0550	1202	1814	12:24	27
Oct 1	0604	1157	1750	11:46	27
Oct 16	0619	1153	1727	11:08	27
Nov 1	0636	1151	1706	10:31	28
Nov 16	0652	1152	1652	10:00	29
Dec 1	0708	1157	1645	9:37	30
Dec 16	0721	1203	1646	9:25	31

Note: Mountain Standard Time in hours and minutes on 24-hour clock. Add one hour during Daylight Time. Aspen, Colorado is Lat. 39°11′N Long. 106°50′W.

Figure 10-2. Bioclimatic index

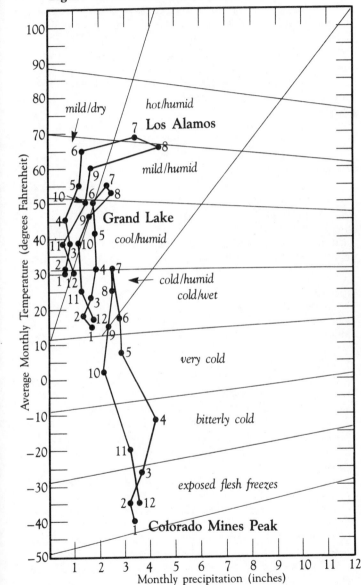

Mines Peak, the winter bioclimate is bitterly cold, primarily because of the exposure to high winds. In the summer, the contrast is not so great but nevertheless significant. Summer bioclimate at Grand Lake can be characterized as on the border between cool/dry and warm/dry; whereas on the summit of the nearby peak, the bioclimate is cold/humid.

Latitudinal variations are less. Los Alamos is only slightly warmer than Grand Lake, particularly when the elevation difference is taken into account. Los Alamos is about 1,200 feet lower than Grand Lake and would be expected to have mean temperatures about 4 degrees warmer. Precipitation is less at Los Alamos (except during July and August), and its bioclimate is predictably warmer and dryer.

During nonstormy periods, therefore, the southern Rockies are suitable for year-round outdoor activities by suitably clothed and equipped recreationists. During winter anywhere in the mountains and especially above the timberline and on exposed ridges and mountaintops, the bioclimate can be severe and life threatening. Winter backcountry skiers must be prepared for bitterly cold and windy conditions.

Summer

Summer is cool, mild, and relatively dry throughout the mountains. Skies are generally clear (87 percent of July days at Alamosa, Colorado, are clear or partly cloudy). From June through September, the sun shines more than 75 percent of the time. This abundance of sunshine coupled with the high altitude means intense solar radiation most of the time and a bioclimate that feels warmer than the modest daytime temperatures indicate.

Air temperature varies with both latitude and altitude. The average July temperature at Aspen, 62°F, is 6 degrees colder than at Los Alamos, 230 miles to the south but at about the same elevation. At Berthoud Pass, the July mean is 51°F, 4 degrees colder than nearby Grand Lake, which is nearly 3,000 feet lower.

The average daily maximum temperature in July at Grand Lake is 75°F; the highest ever recorded was 90°F. At Los Alamos, July temperatures are only slightly higher: the average July maximum is 80°F and the record is 94°F. At Red River, New Mexico, in the heart of the southern Sangre de Cristos, the figures are 76°F and 91°F. Humidities are also low, so the climate is never oppressively hot.

Because of the high altitude and generally clear skies, the diurnal temperature range is large, generally greater than 30 de-

Figure 10-3. Temperature. Aspen, Colorado

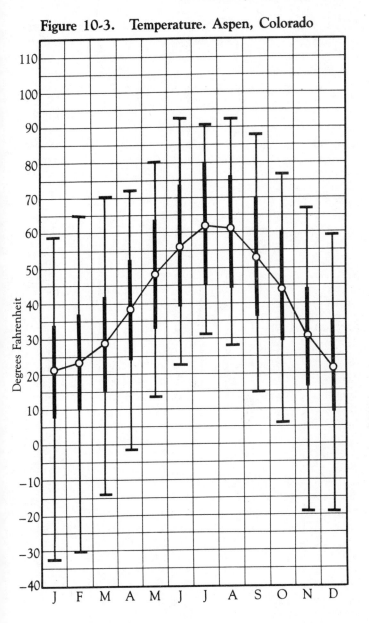

Figure 10-4.   Temperature. Berthoud Pass, Colorado

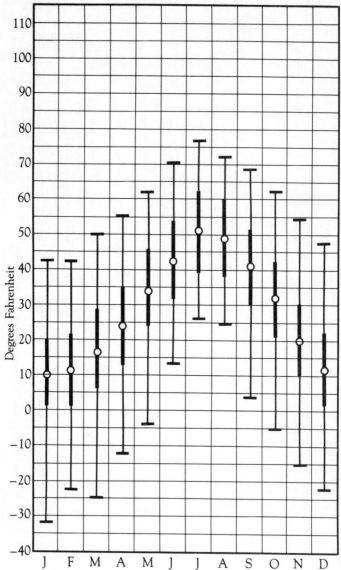

grees and often reaching 40 degrees. Summer nights are therefore rather cool, even cold, with the clear night sky being a massive sink for outward thermal radiation. Cold air drainage adds to this effect, and campers in valley bottom meadows may experience frosts any time during the summer. At Grand Lake, the average July minimum is 36°F and the record low is 21°F. At Red River, the corresponding minima are 40°F and 28°F.

Precipitation is higher in July and August than the rest of the year, although it is still rather light. Nearly all precipitation comes in the form of showers and thundershowers; summer frontal storms are rare. The southern portion of the region receives more summer rainfall than the northern end, primarily because the Bermuda High extends far into the continent and produces a southeasterly flow of moist air toward the mountains. Snowfall is rare during July and August, although late August snows are not unknown. Typically, the first snow of the winter occurs in late September and the last in early June.

Rainy spells lasting two or more days are uncommon, although three such periods may occur in July. Fewer occur in other summer months. Dry spells are the rule: during each month from March through August, one can expect three of them, each lasting three or more days.

Nevertheless, showers are frequent in the mountains. On five to eight days each summer month, precipitation of 0.1 inch or greater can be expected, the greater frequency occurring above the timberline and in the southern end of the range. Heavy rainfalls are rare; generally only one day per summer month will have rainfall in excess of 0.5 inch.

Most of the showers come in thunderstorms, and these pose a distinct hazard to the above-timberline hiker and climber. At Grand Junction, eight days in both July and August will have thunderstorms. At Alamosa, the frequency is higher: more than half the days have thunderstorms (figure 10-6). Lightning frequency is highest on the ridges and summits; prudence suggests hikers attempt to be off exposed places by early afternoon. Summer storms often produce little or no rainfall on the ground. The air is so dry that the rain evaporates completely on the way down. Virga, the streamers of evaporating rain beneath a shower cloud, is a common sight in the Rockies.

Wind is generally rather light in the summer, even above the timberline. Exceptions are on exposed ridges and summits subjected to the free flow of upper-level winds. At Berthoud Pass, above timberline, average wind speed during the summer is about

Figure 10-5. Monthly precipitation

Berthoud Pass, Colorado

Annual: 37 inches   Snowfall: 375 inches

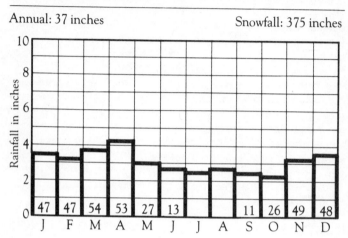

Aspen, Colorado

Annual: 19 inches   Snowfall: 138 inches

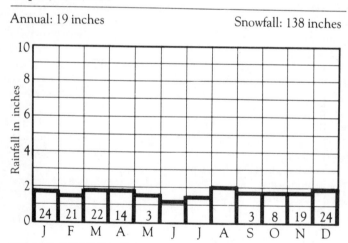

Figures at the base of the columns indicate inches of snowfall

Figure 10-5. *(continued)*

Grand Lake, Colorado

Annual: 20 inches   Snowfall: 156 inches

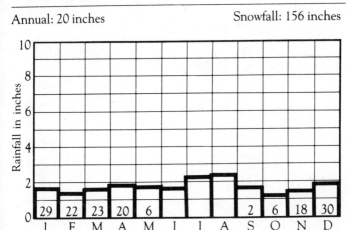

Red River, New Mexico

Annual: 19 inches   Snowfall: 118 inches

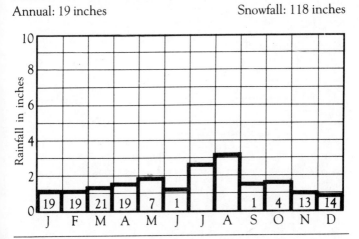

Figures at the base of the columns indicate inches of snowfall.

Figure 10-6. Sky and weather. Grand Junction, Colorado

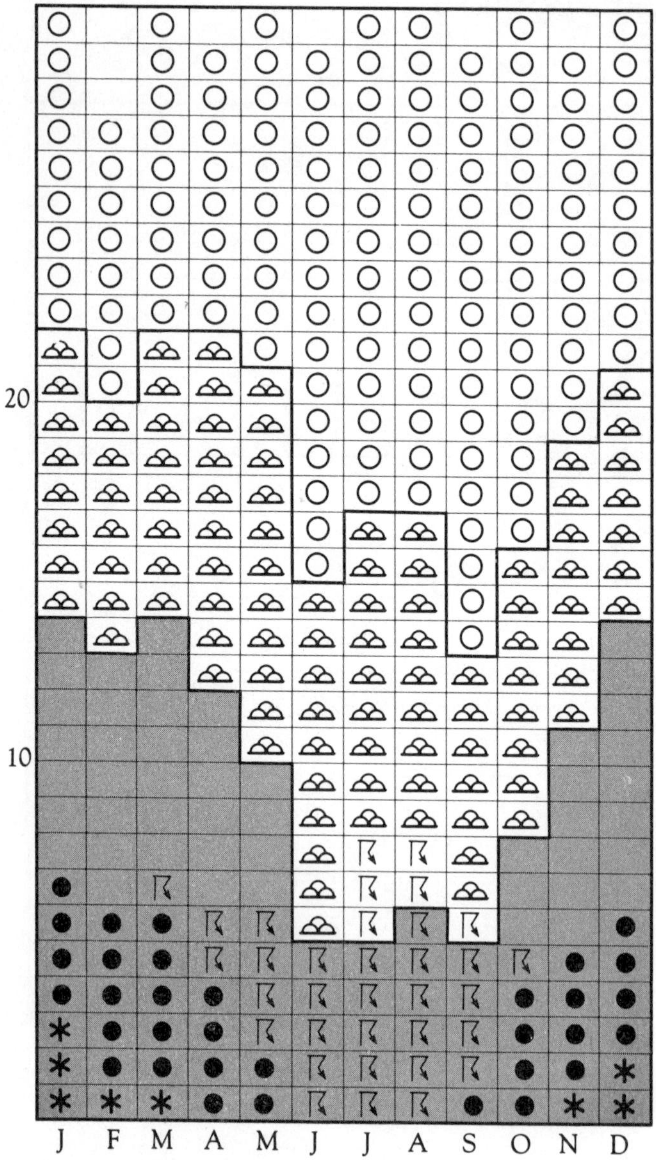

Figure 10-6. (continued) **Alamosa, Colorado**

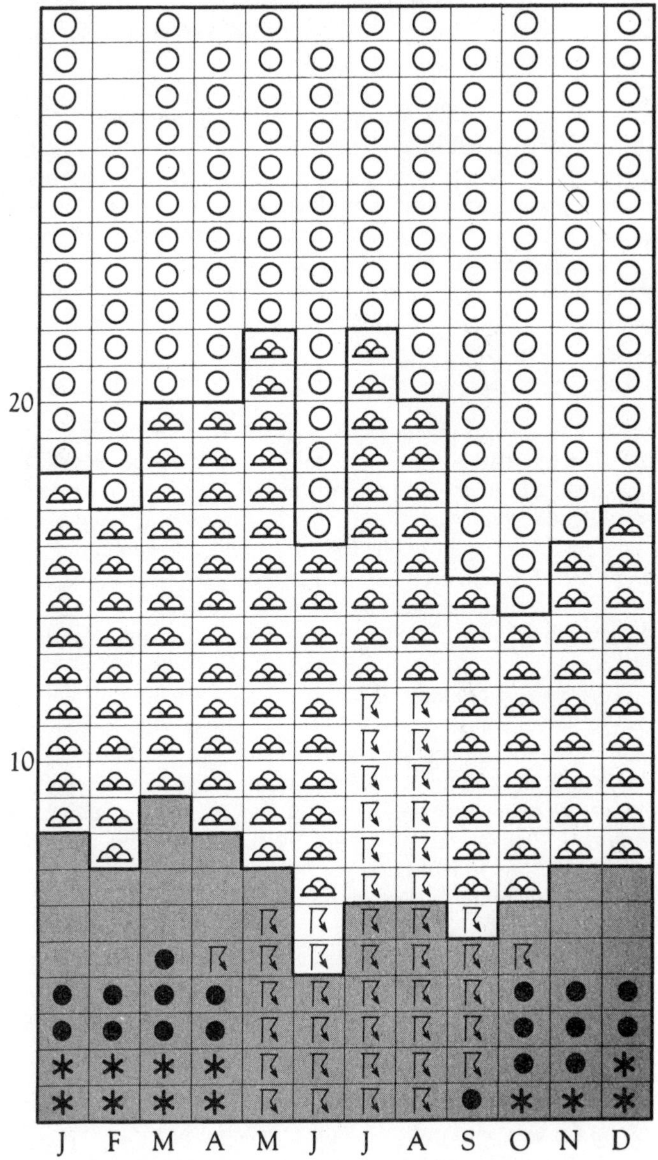

14 mph at the top of a 33-foot tower. This translates to about 10 mph at face level. Below timberline, wind speeds are much lower. Even above the trees, as at Fool Creek (table 10-2), summer winds average only about 5 or 6 mph. Below the canopy, the wind will rarely exceed 1 or 2 mph.

Thus summer in the southern Rockies is nearly ideal for hiking. Whenever trails are open, usually by the end of June, until the first snowfall in late September, conditions are good for backpacking. Days are typically sunny and cool, with afternoon showers common. Humidities are low; nights are cool and generally rainless. The frequent thunderstorms pose the major hazard. Hypothermia weather is rare in the summer, although rain coupled with low temperatures and strong winds can occur at either end of the summer season.

Autumn

Autumn is short in the southern Rockies. The aspens turn in September, splashing yellow and gold across the high slopes. By the end of September, certainly by the end of October, the snows of winter take over the mountains. The minimum temperature goes below freezing on about half the nights in September, on most of the nights in October, and on every night in November. Daytime maximum temperatures also show a rapid decline from highs averaging near 70°F in September to the mid-forties in November.

Precipitation declines also, but most of it now appears as snow. At lower elevations and in the southern region, rain may persist through October. By November it is generally very light and powdery snow. Thunderstorms diminish through September, although there may be the occasional one in October, about one per month. Avalanches are generally not a problem until the snowpack is 20 to 30 inches deep.

With the decline in seasonal precipitation comes a period of bright, sunny weather. In September and October, more than 50 percent of the days are clear; fewer than 20 percent are classed as cloudy. Dry spells of four or more days occur about twice each month. Only about once per month is there a three-day stormy period.

October is not a particularly windy month; the average is generally less than that in September or November.

The transition month of October and the just-previous week or two in September are chancy for either hiking or skiing. There

Table 10-2.   Mean and Maximum Wind Speed

Elevation of anemometer	Fool Creek, Colorado 10,620' 60'	Berthoud Pass, Colorado 11,880' 33'	Mines Peak, Colorado 12,493' 38'
January	11 (47)	17 (66)	34 (100)
February	8 (37)	15 (54)	29 (85)
March	9 (45)	15 (56)	28 (77)
April	9 (38)	15 (50)	23 (71)
May	8 (34)	13 (47)	18 (57)
June	7 (30)	14 (48)	25 (59)
July	5 (40)	13 (53)	21 (77)
August	6 (25)	14 (45)	24 (50)
September	7 (40)	14 (53)	24 (77)
October	6 (40)	14 (53)	23 (77)
November	7 (50)	14 (50)	27 (88)
December	7 (46)	15 (56)	30 (90)

Note: Based on eight to twelve years of data. Maximum speeds in parentheses are for one-hour average. Values for June through October at Berthoud Pass and Colorado Mines Peak are estimated from Fool Creek data.

are exceptions to this, of course; occasional Octobers will have little or no snow, even at high elevations. Backcountry travelers are well advised to pay particular attention to weather signs and forecasts and always be prepared for cold, snowy weather.

Winter

Defined in terms of those months when backcountry travel must be by ski or snowshoe, winter lasts from November through May and occasionally into June and even July. The mean snow depth at Berthoud Pass on June 15 is 12 inches, but snow usually disappears by the end of the month.

January is the coldest month, averaging 10°F at Berthoud Pass and 10 degrees higher in the valleys. Farther south, at Mesa Verde National Park and Los Alamos, New Mexico, the January average is 30°F. December and February are only about 3 degrees warmer. March sees a substantial warming trend, but the average is still below freezing everywhere but in the extreme southern portion of the region. Maximum temperatures rise above the freezing mark most days in March, aided by the strong sunshine. One-third of March days are clear; another third are only partly cloudy.

Midwinter precipitation is light, averaging 1 to 2 inches of water equivalent per month, all of which comes as light, powdery snow. At elevations above the timberline, amounts may be double these figures, totaling 4 feet of powder snow per month. At Berthoud Pass, although the total annual precipitation is only 37 inches, the snowfall averages 375 inches. In the valleys, snowfall totals about half this amount. The snow tends to come in relatively frequent storms that drop light to moderate amounts of snow. At Berthoud Pass, an average of eleven storms per winter month produce more than 0.1 inch of precipitation; but only one produces more than 0.5 inch (table 10-3). It is a seeming paradox that such small amounts of precipitation produce such large amounts of snow. The answer is in the extremely low density of the snow. This is truly powder-snow country.

Winds are somewhat higher in the winter than in other seasons. The highest monthly average winds occur in January: 34 mph at the top of Colorado Mines Peak; 17 mph at Berthoud Pass; and 11 mph at Fool Creek, just below the timberline. These high winds combine with low winter temperatures to produce a severe above-timberline bioclimate. November through April all have average conditions above the timberline that are bitterly cold.

This combination of frequent snows accumulating to great depths, alternating temperature, and high wind produces some of the worst avalanche conditions found in the mountains of North America. Colorado leads all states in the number of avalanche fatalities. About 90 percent of the avalanches in the southern Rockies occur during or just after snowstorms. Weather patterns contributing to snow avalanches include large quantities of cold, new snow accompanied by strong winds; rain falling on cold, fresh snow; and thaws that warm the snowpack to the melting point. Backcountry skiers, even ski area users, should constantly be aware of and prepared for the danger of avalanches. The National Weather Service and the U.S. Forest Service operate an avalanche warning service in Colorado. More information on avalanche weather is contained in chapter 5.

Spring

Spring, as we have defined it, is almost nonexistent. Temperatures climb rapidly through April and May to June. But heavy snowfalls can and do occur in April and May. The heaviest snow-

Table 10-3. Average Number of Days per Month with Precipitation Equal to or Greater than Indicated Amounts

Month	Grand Lake 0.1''	0.5''	Berthoud Pass 0.1''	0.5''	Aspen 0.1''	0.5''	Los Alamos 0.1''	0.5''
January	6	0	11	1	6	0	2	0
February	5	0	11	1	6	0	2	0
March	6	0	12	2	5	1	3	0
April	7	1	11	2	6	1	2	0
May	6	0	8	1	4	1	3	0
June	5	1	7	1	4	0	3	1
July	6	1	8	1	5	0	8	2
August	7	1	8	1	6	1	9	3
September	5	1	7	1	5	1	4	1
October	3	0	6	1	4	1	3	1
November	4	0	10	2	5	0	2	0
December	6	1	11	1	6	1	3	0
Annual	66	6	110	15	62	7	44	8

fall ever recorded at Berthoud Pass occurred during a late April snowstorm in 1933. In seventy-two hours, 94 inches of new snow fell. And the largest twenty-four-hour snowfall recorded in the United States occurred nearby in April 1922, when 76 inches fell. Normally, the snowpack reaches its maximum in April, then declines steadily, disappearing by mid-June.

The biggest jump in temperatures occurs from April to May, about a 10-degree increase in the mean. Precipitation totals remain about the same as in the winter months, but there is a shift to more rain and less snow. Snow density also increases during April and May storms, reaching a soggy 17 percent in June. Days become long, and the sun reaches high in the sky. Nevertheless, a bright sun does not guarantee comfortable skiing. I remember well one April ski tour along Keystone Ridge: a brilliant sunny day and ideal snow conditions. But on the exposed ridge, a cold northwest wind was blowing at about 40 mph, making progress difficult. Below the ridge, in the trees and out of the wind, it was shirtsleeve weather.

Winter can be said to end and spring to begin in May in protected places. The transition from winter to summer comes at different times in different microclimates. Once the snow is gone, it is really summer everywhere.

Summary

The relatively mild continental climate of the southern Rockies produces some of the best outdoor recreation climate to be found anywhere in the United States and Canada. Summer months are warm and dry although showers and thunderstorms are frequent. But total rainfall amounts are low, averaging about 2 inches per month.

The transition periods of spring and fall are almost non–existent. The coming of winter is frequently heralded by a late September snowstorm and the winter season's last snowstorm is usually in May. The snowpack lasts until late June or even into July.

Winter in the southern Rockies offers excellent snow-based recreation. A long snow season, with deep snowpacks lasting until May or June, normally permits downhill and backcountry skiing for five or six months. Frequent short-duration storms alternate with sunny periods of several days' duration: ideal for ski touring but also ideal for avalanche formation. The weatherwise skier who pays attention to official forecasts and local signs and understands safe route finding through potential avalanche areas will be able to avoid nearly all danger.

Additional information

Two Forest Service publications (available from the Rocky Mountain Forest and Range Experiment Station, USDA Forest Service, Ft. Collins, CO 89521) contain a wealth of information on the high altitude climate of the Colorado Rockies. They are:

Judson, Arthur. *The Weather and Climate of a High Mountain Pass in the Colorado Rockies.* USDA Forest Service Research Paper RM-16. 1965.

Judson, Arthur. *Climatological Data from the Berthoud Pass Area of Colorado.* USDA Forest Service General Technical Report RM-42. 1977.

For the southern end of the region, the following report (available from the National Climatic Center, Federal Building, Asheville, NC 28801) is useful:

Red River, New Mexico. Climatic Summaries of Resort Areas, Climatography of the United States No. 21-29-1.

Chapter 11

The Cascades

The Cascades began some 200 million years ago when the North American continent and the Pacific Ocean floor started to push against each other. When ocean sediment and continent collided, the heavier ocean plates started to slide under the continent. As the sediments piled up against the continental edge, they pushed up a range of mountains along the coast. Farther inland, a string of volcanoes formed more or less parallel to the coast as melted basalt from the ocean plate pushed its way through the surface.

The present Coast Range and the inland Cascades represent a second cycle of this sequence. The remnants of the first coast ranges can be seen in the Blue Mountains in eastern Oregon. Today's pattern started about 35 million years ago when the seafloor sank along a line parallel to the present coastline. A period of intense volcanic activity followed, spattering basalt and volcanic ash over much of what is now Washington and Oregon. There have been other locational shifts of volcanic activity since then; the present volcanoes strung along the Cascades have developed only in the last million years or so. The geological picture is immensely complicated.

From a climatological point of view, the importance of this hastily sketched picture lies in the presence of two mountain chains parallel to the coast, oriented more or less perpendicular to

Figure 11-1. The Cascades

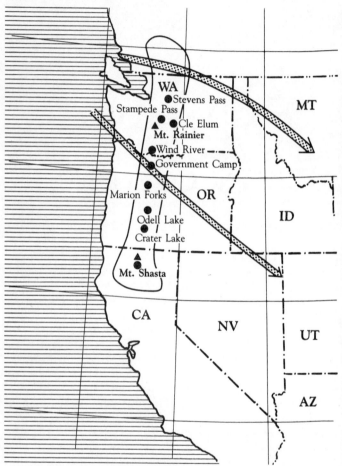

Arrows indicate winter storm tracks.

the prevailing ocean winds. Although the Coast Range is relatively low, it manages to extract prodigious quantities of water from the damp marine air forced over it. Near the shore, the annual rainfall ranges from about 80 inches to 120 inches or more along the coastal mountain slopes. In northern California, this is where the stately redwood grows. Farther north, Douglas fir, Sitka spruce, and western hemlock take advantage of the high rainfall and mild climate to grow to enormous size. Although its trees are not as tall as the redwood or as great in diameter as the giant sequoia, the north coast forest is impressively vast.

East of the coast mountains, there is a marked rain shadow. In the Willamette Valley, for example, the annual precipitation is about 40 inches, not exactly a desert but much less than the 120 inches just a few miles west. Continuing east the Pacific air is forced over a second range—the much higher Cascades. Many of the peaks along the main crest are more than 10,000 feet high; the highest volcanic peaks tower more than 14,000 feet. Once again, precipitation is wrung from the ascending air masses. Much of this falls as snow at the higher elevations and feeds the more than nine hundred glaciers that clothe many of the peaks. The precipitation also feeds the thousands of lakes scattered throughout the Cascades.

This double depletion—first the Coast Range, then the Cascades—robs the Pacific air of most of its moisture. East of the mountains lies a near-desert, with annual precipitation of 12 inches or less.

General climate

The major meteorological control over the Cascades climate is the relative position of the two great semipermanent pressure systems in the northern Pacific: the Aleutian Low and the North Pacific High. In the winter, the Aleutian Low intensifies and spins off traveling lows that bring frontal storms to the Washington and Oregon coasts. The main weather storm tracks are thus from the northwest to the southeast across the Pacific Northwest area. However, the circulation around these lows is such that the air masses arriving at the coast have a long trajectory over relatively warm water. They arrive with a high moisture content. Because of the counterclockwise circulation around the lows, the rain generally arrives on southerly winds. Well above the surface, the wind stream is more nearly from the southwest or west.

Fronts associated with these winter storms are often rather indistinct, the frontal wave having closed up or occluded offshore. One storm frequently succeeds another in close succession, so they merge with one another. Coastal cities have a well-deserved reputation for being rainy in the winter: about two-thirds of winter days have measurable precipitation.

Occasionally, a strong low-pressure area migrating eastward across the southern part of the region will bring north winds behind it, flooding the area with cold Canadian air. Although the region east of the Cascades is typically occupied by a dry continental high-pressure area, the mountains act as an effective barrier. Only rarely does this cold, dry air cross the mountains; when it does, coastal cities record their lowest temperatures.

In the summer, the Aleutian Low migrates northward and the region is dominated by an intensified North Pacific High. The clockwise circulation brings air from the northwest. Because water temperatures are low over the North Pacific, the air is relatively dry. The warm land surface triggers vertical convection; afternoon cumulus clouds are common. Showers and thunderstorms occur fairly often in the mountains, but some of the thunderstorms do not produce enough rain to reach the ground. These dry lightning storms are especially feared by the foresters, for they often start backcountry forest fires.

The annual precipitation cycle is thus somewhat similar to that farther south in California, where a typical Mediterranean climate prevails: wet winters and dry summers. The difference is in the strong moderating influence of the marine air: summer and winter temperatures are mild. In many ways, the climate is more like that of western Europe, with its climatic moderator, the Gulf Stream. Here the similarity ends, for the European mountains are oriented east-west, permitting deep penetration of marine air into the continent. Our west coast mountains stretch north-south and prevent marine influence from reaching eastward beyond their crest.

There is also a rainfall gradient from north to south. In the northern Cascades, annual precipitation is generally over 100 inches; in the Oregon Cascades, annual amounts are generally less than 72 inches; and in the southern extremities in northern California, the totals are 50 inches or less.

The mountains exert a tremendous control over the climate. A west-to-east transect across Oregon, along the forty-fifth parallel just south of Portland, starts out with 72 inches of annual precipitation on the coast. At the crest of the Coast Range, it reaches 120 inches, then drops to 40 inches in the Willamette Valley. Near the crest of the Cascades, the value is up to 7 inches, then drops rapidly to 12 inches at the eastern foot of the mountains. These changes take place in a horizontal distance of 125 miles.

Solar data

Data for Government Camp, Oregon (which is representative of the Cascades) are presented in Table 11-1. Because the region extends nearly due north-south, daylength varies markedly. In the north end, near the U.S.-Canadian border, winter days are 40 minutes shorter and summer days about 45 minutes longer than those at Government Camp.

Bioclimatic index

Because of the strong topographic controls over the climate of the Cascades, the region has a number of typical bioclimates. For comparison, we can look at climatic records from Stampede Pass on top of the main divide of the Cascades (elevation 3,958 feet) and from Cle Elum, 20 miles east (elevation 1,930 feet). Stampede Pass is the highest regular reporting station along the crest and will be typical of much of the lower part of the high country. Cle Elum is 20 miles east of the divide, 2,000 feet lower, and very much in the rain shadow of the main range.

The bioclimate of Stampede Pass ranges from cold/wet in the winter (Novermber–March) to mild/humid in the summer (June–September), except for July, which just falls into the mild/dry zone. Based on afternoon relative humidities that average near 60 percent, it is likely that even July should be in the mild/humid zone. At higher elevations, the curve would be displaced downward, about 15–20 degrees for peaks near 10,000 feet. Higher wind speeds in this above-timberline zone also indicate a colder bioclimate, and it is probable that winter conditions at high elevations would be in zone II (very cold) or zone III (bitterly cold). Cold/humid conditions prevail above the timberline in the summer months.

At Cle Elum, because of the rain shadow effect, the curve is displaced leftward and slightly upward; so the winter months are predominantly cold/humid. Because of substantially warmer summertime temperatures, May through September are classed as mild/dry.

At the town of Mt. Shasta, which is nearly the same elevation as Stampede Pass but in the very southern end of the Cascades, summer months are very much like those at Cle Elum:

Table 11-1.   Solar data for Government Camp, Oregon

Date	Sunrise	Solar noon	Sunset	Day length hr:min	Twilight min
Jan 1	0746	1210	1634	8:48	34
Jan 16	0742	1217	1651	9:09	33
Feb 1	0728	1221	1713	9:45	32
Feb 16	0708	1221	1735	10:27	31
Mar 1	0644	1219	1754	11:10	30
Mar 16	0617	1216	1814	11:58	30
Apr 1	0546	1211	1835	12:49	30
Apr 16	0519	1207	1854	13:35	32
May 1	0455	1204	1913	14:19	33
May 16	0435	1203	1932	14:56	36
June 1	0422	1205	1948	15:26	38
June 16	0418	1208	1957	15:39	39
July 1	0423	1211	1959	15:35	38
July 16	0435	1213	1951	15:16	37
Aug 1	0452	1213	1934	14:42	35
Aug 16	0510	1211	1912	14:02	32
Setp 1	0529	1207	1844	13:15	31
Sept 16	0547	1202	1816	12:28	30
Oct 1	0606	1156	1747	11:41	30
Oct 16	0625	1152	1720	10:55	30
Nov 1	0647	1151	1654	10:07	31
Nov 16	0708	1152	1636	9:28	33
Dec 1	0727	1156	1626	8:59	34
Dec 16	0741	1203	1625	8:44	35

Note: Pacific Standard Time in hours and minutes on 24-hour clock. Add one hour during Daylight Time. Government Camp, Oregon is Lat. 45°18′N Long. 121°45′W.

mild/dry. The winter months are substantially warmer than either Cle Elum or Stampede Pass, but intermediate in precipitation, putting those months near the boundary between cool/humid and cool/wet. Although the curve is not shown on the bioclimatic diagram, it would occupy the space between the Cle Elum and Stampede Pass curves.

The bioclimate then is relatively equable, though somewhat humid throughout the year. Generally comfortable conditions

Figure 11-2. Bioclimatic index

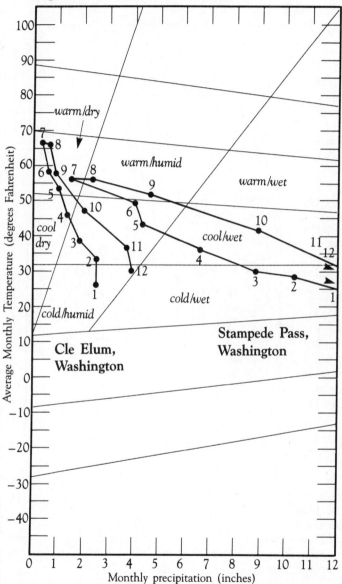

Figure 11-3.   Temperature. Stampede Pass, Washington

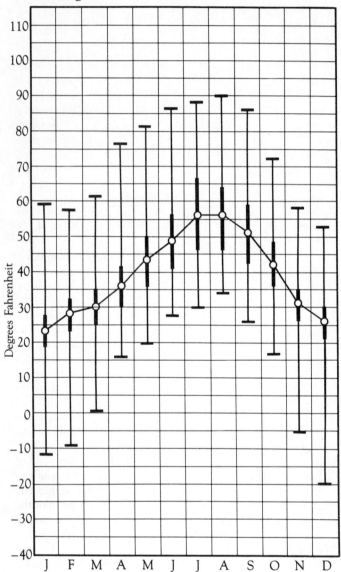

prevail in the summer, with the highest temperatures associated with low rainfall, especially in southern portions. Even in winter, elevations below the timberline have moderate conditions. Only above the timberline are conditions likely to be severe.

Summer

Summer arrives late but lasts rather long at middle elevations in the Cascades. The mean temperature normally reaches above 50°F late in June and drops below 50°F in late September at Stampede Pass (elevation 3,958 feet). Daily maximum temperatures are generally in the mid-sixties in July and August, dropping to near 50°F in June and September. At higher elevations, these figures will be somewhat lower. At Paradise Ranger Station near the timberline on Mount Rainier (elevation 5,550 feet), temperatures are 3 or 4 degrees lower.

Nighttime minimum temperatures at mid-elevations are generally in the mid-forties. Above timberline, frosts are not uncommon any time during the summer but are rare below 4,000 feet.

At Crater Lake National Park headquarters (elevation 6,500 feet), eight or nine days have maximum temperatures above 75°F; only two or three days have minimum temperatures below freezing during July and August. Temperatures are thus rather mild, even cool, in the summer. The greatest range between daytime maximums and nighttime minimums occurs when winds are easterly. These winds bring dry air and clear skies from the interior.

Summer days tend to be clear or partly cloudy, although about one-third of the days are predominantly overcast in the northern portion of the region. The number of clear days increases as one goes south; most days are clear in the vicinity of Mount Shasta in midsummer. In the extreme south, only about three days are cloudy during July, August, and September. This portion of the region is very much like the Sierra Nevada.

Summer rainfall is also more frequent in the northern end of the Cascades—about ten days per month—than in the south— three or four days per month. Thunderstorms are infrequent throughout the region, averaging about two or three per month. But in the southern end, nearly all summer precipitation comes from thundershowers; whereas in the north, most of the rain is not accompanied by thunder.

There is also a considerable north-south gradient in the rainfall totals. July and August average about 2 to 3 inches each in the north, but less than 1 inch each month in the south.

Figure 11-4. Sky and weather. Stampede Pass, Washington

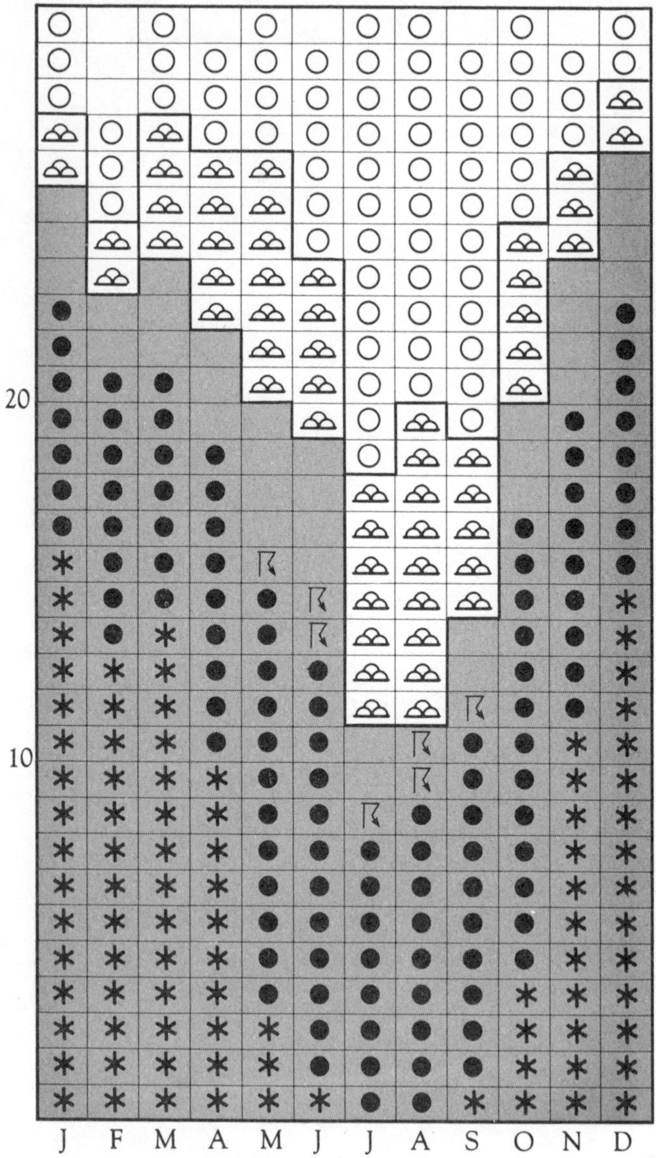

Figure 11-4. *(continued)* **Mount Shasta, California**

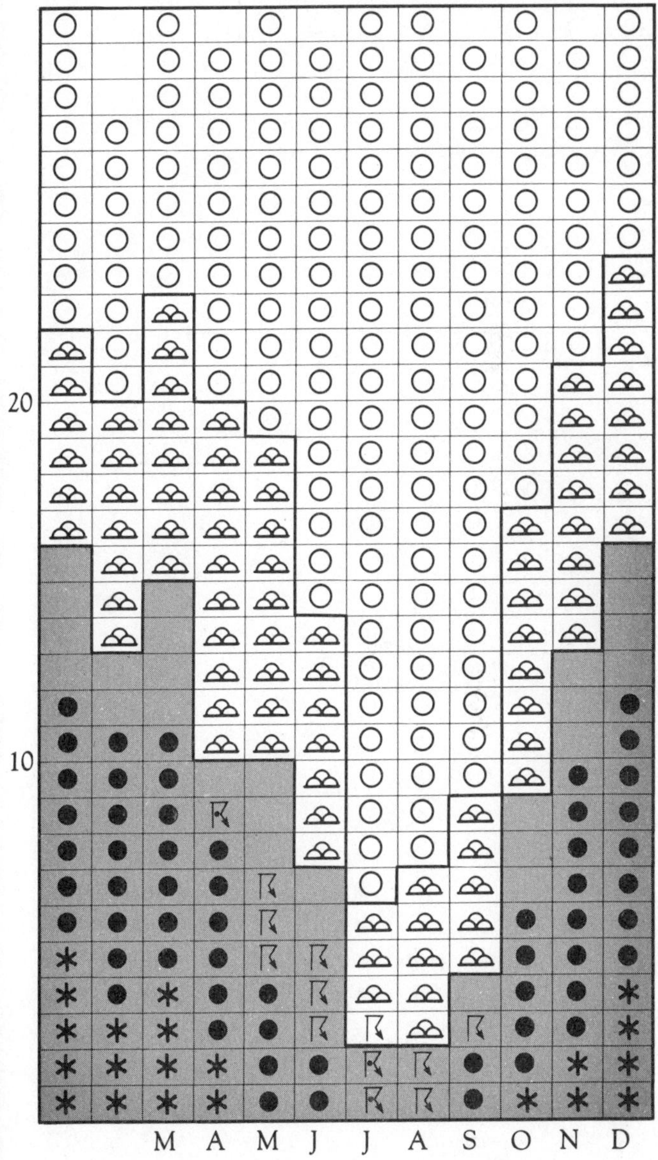

At the north end of the Cascades, in Washington, the best hiking times are from July through October, although above the timberline the snow usually starts to fly by the end of October. Daytime temperatures are moderate; nights are cool. Sunny or partly sunny days predominate, but rain protection is essential, especially in the North Cascades. Above-timberline passes are typically snowed in until the end of June.

Wind speeds tend to be light to moderate, even in exposed places.

Figure 11-5.   Precipitation frequency. Government Camp, Oregon

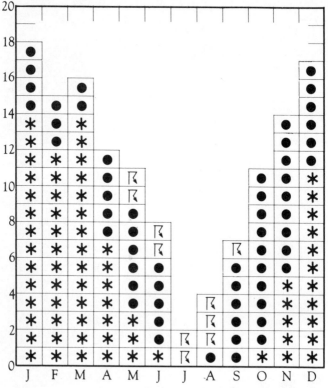

Number of days with 0.1 inches precipitation or more. Snow symbol indicates days with one inch of snow or more. Closed circles indicate rain.

Figure 11-6. Monthly precipitation

Cle Elum, Washington

Annual: 22 inches

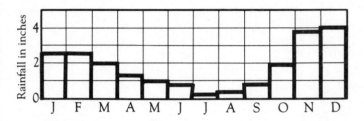

Stampede Pass, Washington

Annual: 91 inches   Snowfall: 445 inches

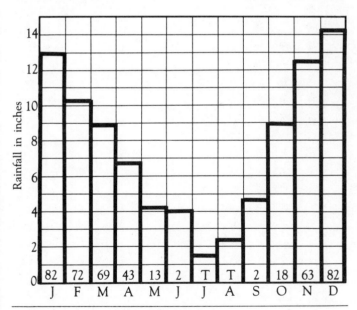

Figures at the base of the columns indicate inches of snowfall.

Figure 11-6. *(continued)*

Mount Shasta, California

Annual: 37 inches   Snowfall: 110 inches

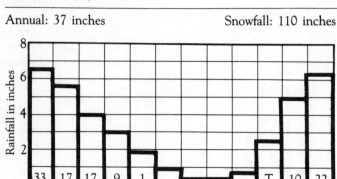

Crater Lake National Park

Annual: 67 inches   Snowfall: 540 inches

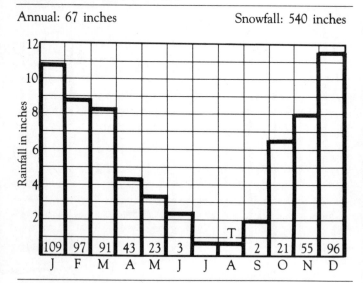

Figures at the base of the columns indicate inches of snowfall.

Autumn

The transition between the relatively rain-free summer and the heavy precipitation of fall and winter usually begins in October. At higher elevations, fall precipitation is frequently in the form of snow, although periods of snowfall are often followed by heavy rains, especially below 8,000 feet. Lowland flooding may occur under such conditions, and streams may be difficult or dangerous to ford.

Temperatures decrease rather rapidly during October and November; minimums below freezing are typical at mid-elevations by the first of November. However, the fall daily range is very small and daytime maximums may not rise much above freezing. Most days are cloudy and rainy. Even in the southern portion, only a third of the days are clear.

Autumn is not very propitious for outdoor activities: frequent cold rains or wet snow, with predominantly cloudy skies. Conditions are somewhat better in the southern portion, in the Trinity Alps and the Mount Shasta area in California. But the hiker must be prepared for damp, cold weather in the fall season anywhere in the Cascades.

Winter

By the end of October at high elevations and the end of November below 4,000 feet, the Cascades are in the full grip of winter. This is not so much a matter of extremely low temperatures—the December mean is 26°F and the January mean only 23°F—but of the incredible amounts of snow that blanket the mountains. At low elevations, winter season snowfall ranges between 50 and 75 inches. The amount increases with elevation, reaching 400 to 600 inches at elevations of 4,000 to 5,500 feet. It is probable that maximum snowfall amounts occur at about 7,000 feet, but there are no recording stations at high elevations. It is certain that winter snowfall amounts of 1,000 inches have occurred: in the winter of 1955–56, this amount was recorded at Paradise Ranger Station on Mount Rainier. The greatest depth of snow on the ground recorded there was 367 inches, more than 30 feet, in March 1956.

Snowfall amounts decrease southward but are nevertheless astounding. At Crater Lake, a maximum depth of snow on the ground reached 242 inches in 1927; the greatest seasonal total was 879 inches in the winter of 1932–33.

Figure 11-7. Probability of first and last one-inch snowfall. Northern Cascades

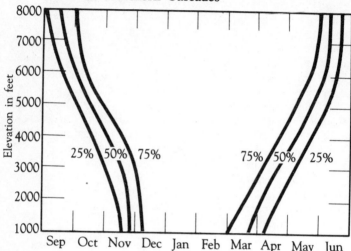

Snowfall and snow depth depend on elevation. Figures 11-7 and 11-8 show the probability of snowfall and snow depth as a function of elevation. Although the diagrams were prepared for the Oregon Cascades, they will be nearly correct for the northern portions of the range as well.

Figure 11-7 indicates the probability (expressed in percentage figures) that a 1-inch snowfall will occur as early or as late as the dates indicated by the appropriate curve, as a function of elevation. For example, there is a 25-percent probability that the first 1-inch snowfall will occur at 6,000 feet on or before September 19. The probability is 75 percent that a 1-inch snowfall will occur at that elevation by October 14. In the spring, the probability that a 1-inch snowfall will occur on or after June 14 is 25 percent.

Figure 11-8 gives the probabilities of at least a 6-inch snow depth on the ground occurring between the indicated dates. Thus at 6,000 feet, there is a 25 percent probability that there will be 6 inches or more of snow on the ground by October 10; that is, in one year out of four, the depth will be at least 6 inches by that date. Figure 11-8 also gives the probabilities of 6 inches or more remaining on the ground in the spring. Thus there is a 75-percent chance that there will be at least 6 inches of snow on the ground

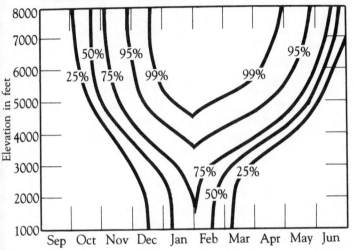

Figure 11-8. Probability of six-inch snowpack on indicated date. Northern Cascades

at 6,000 feet on June 4. Figure 11-8 indicates that there is better than 95-percent probability that the snow depth will be at least 6 inches at 5,000 feet between December 11 and April 15 and that snow depths of 6 inches do not occur every year below about 4,500 feet. At 2,000 feet, only in three winters out of four will the snow depth reach 6 inches.

As might be expected from the relatively warm winter temperatures, Cascades snow tends to be wet and heavy. This copious wet snow is just right for keeping the glaciers going; but it is also just right for producing wet-snow avalanches. The most disastrous avalanche in American history swept ninety-six people to their deaths near Steven Pass in March 1910. Two railroad trains were stalled by small avalanches that had buried the tracks. Passengers were waiting out the storm in the presumed safety of the stalled trains when a massive avalanche swept many of the cars off the tracks and down a precipitous slope. All but twenty-two of the passengers and crew perished.

A compilation of avalanche accidents in the United States between 1967 and 1971 shows twenty-one out of seventy-six occurred in the Cascades of Washington. Curiously, only one of the seventy-six occurred in Oregon, on Mount Hood, although this may reflect in part the heavier winter use of the northern Cascades.

With all the winter snow, it might be expected that the sun never shines on the Cascades. That is very nearly true; only about three days each winter month qualify as clear, and only about three more qualify as partly cloudy (figure 11-4). The rest are cloudy and it snows (or rains) on almost all of them. Even in the sunny Southern end of the range, half the days are cloudy; about one-third have some precipitation.

Winter winds are generally moderate, but winds of hurricane force have been recorded every winter at coastal stations. Farther inland, wind speeds are diminished by the retarding effect of the rough mountainous surface. Above the timberline, winds of moderate speeds can be expected as a matter of course during storms.

In summary, winter in the Cascades is a time for snow lovers. Excellent ski mountaineering conditions prevail, but wilderness touring is not for the novice. Backcountry touring requires protection against the almost daily wet snowfall and a good knowledge of avalanche conditions. It is certainly not for one who demands bright, sunny weather.

Spring

Spring is a time of gradual drying out of the atmosphere and a melting of the immense snowpack. In late May, there is still a 50-percent probability of a snowfall of at least 1 inch. And there is a 50-percent probability that there will be at least 6 inches of snow on the ground. June is the transition month above the timberline; spring comes a month or so earlier at lower elevations. Study of the snow probability diagrams will help clarify the time-elevation relationships.

The high passes will probably still have snow through June and occasionally into July. The snow (except for the glaciers) is gone from even the highest areas by mid-August. High-country hikers should check the previous winter's snowfall amounts before planning a June foray.

Summary

The recreation climate of the Cascades is dominated by three major factors: their great range in altitude, their long latitudinal reach, and their position athwart the currents of moist air from the Pacific Ocean. These work together in winter to produce tremendous amounts of heavy snow at high elevations in the north end of the range, lesser amounts at low elevation and at the south end. Above five thousand feet, snow lies deep nearly every winter. Below three thousand feet, many winters will see only a sparse snow cover. Finding good winter recreation is then primarily a matter of going high enough. Precipitation—rain or snow—is frequent.

A marked reversal occurs in the summer. Rainfall amounts are light, skies are sunny, and air temperatures high, especially in the southern end of the range. Good hiking weather prevails for the period from early July to late September or early October. The western side of the crest has more rain—perhaps four inches per month—than the arid east side. Summer rainfall amounts are greater in the north than in the south. Temperatures are generally mild and a summer-weight sleeping bag is usually adequate except at elevations above timberline. Good protection from rain is essential except, perhaps, on the drier east side. Above timberline, winds can be uncomfortably high and temperatures low enough to require cold-weather clothing.

Additional information

There appear to be few good climatic summaries of Cascade climate. Two with useful mountain climate data are available from the National Climatic Center, Federal Building, Asheville, NC 28801:

Upper Cascades of Oregon. Climatic Summaries of Resort Areas. Climatography of the United States No. 21-35-1. U.S. Dept. of Commerce, Environmental Science Services Administration. 1968.

Mount Rainier National Park, Washington. Climatography of the United States 20-45. U.S. Dept. of Commerce, Weather Bureau. 1960.

Mountaineering, the Freedom of the Hills, edited by Peggy Ferber and published by The Mountaineers, Seattle, Washington, has a chapter on mountain weather with superb photographs of clouds over the Cascades.

Figure 12-1. The Olympic Peninsula

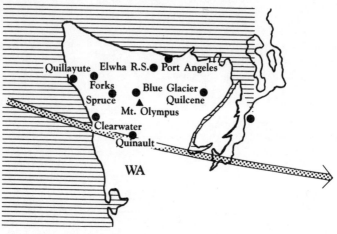

Arrow indicates winter storm track.

Chapter 12

The Olympic Peninsula

In the far northwest corner of the contiguous United States lie the Olympic Mountains. They occupy the Olympic Peninsula, which is surrounded by the Pacific Ocean to the west, the Strait of San Juan de Fuca to the north, and Puget Sound to the east. The mountains are rugged, but scarcely as spectacular as the massive Cascade peaks a few miles to the east. Mount Olympus, the highest, is 35 feet shy of 8,000 feet. One hundred miles to the southeast lies Mount Rainier, rising 14,408 feet.

An extraordinary combination of wind, water, and topography on the Olympic Peninsula has produced a true temperate rain forest in the lowlands and the heaviest rainfall amounts in the continental United States. Annual precipitation ranges from 70 to 100 inches along the coastal plain on the west side of the mountains to more than 150 inches along the west-facing slopes. At low elevations (below 1,500 feet), this is nearly all rain. As a result, the valleys and coastal plain support an extraordinarily lush vegetation. Towering Douglas fir, Sitka spruce, and western hemlock crowd each other for a place to grow. The heights of many of these giants exceed 200 feet. The needles and branches intercept much of the rainfall and continue to shed it after the rain has stopped. In the heart of the forest, it is often difficult to know whether it is raining above the canopy.

Shrubs and giant ferns on the forest floor respond in lush profusion to the steady input of life-giving moisture. A fallen tree quickly becomes covered with spongy mosses and the processes of decay and return to the earth proceed rapidly. Except for the cool temperatures, casual visitors might well think they are deep within a tropical rain forest.

Only a few thousand feet above, the rain forest gives way to vegetation more typical of the Pacific Coast mountains: western white pine and true firs. Higher up, alpine fir and mountain hemlock dominate, then become dwarfed and disappear at the tree line (about 5,000 feet). Above this level lies the true arctic tundra, stretching upward to the mountaintops. The summit of Mount Olympus (7,965 feet) is scarcely 35 miles from the Pacific Ocean, and one can traverse the life zones from rain forest to tundra in just a few miles. From the terminus of the Hoh River road at the boundary of Olympic National Park, the hiker can walk through typical rain forest, passing the largest Sitka spruce in the park—more than 13 feet in diameter at breast height. Climbing upward to 5,000 feet, about 17 miles from the beginning of the trail, the terminus of Blue Glacier is reached just at the timberline. Above this level, all is tundra and ice or snow.

But the Olympic Peninsula holds yet more surprises, for the topography that extracts so much moisture on its windward side deprives the leeward side of that moisture. Port Angeles, on the north coast, receives only 25 inches of rainfall per year; the windward slopes, scarcely 25 miles distant, receive 150 inches.

The mountains interact with the moisture-bearing winds to produce a complicated climate pattern, from humid, temperate rain forest to alpine glaciers to semiarid temperate forest all within a few miles. To understand these variations, we must look at the major climatic controls.

General climate

There are three major controls over the climate of the Olympic Peninsula: the Pacific Ocean to the west, the wind pattern over the peninsula, and the terrain.

The northwest coast has an archetypical temperate marine climate. The Japanese current keeps the coastal waters relatively warm in the winter and cool in the summer. Water temperatures along the Pacific and the Strait of Juan de Fuca range from 45°F in February to 55°F in August, and the air in its long trajectory over these waters adopts these temperatures.

The general circulation over the peninsula is controlled by the strength and position of the Pacific High and the Aleutian Low. In the summer, the Pacific High is displaced northward and causes a flow of air from the northwest. At this season, the water along the coast is cooler than that farther west; the air is cooled and stabilized as it approaches the coast. Because the water is cool, evaporation from it into the air is low and the approaching air mass in late spring is comparatively dry. Warmed by contact with the ground as it moves inland, the air towers upward in swelling cumulus clouds that release their moisture in summer showers. Nevertheless, the summer air is drier than the winter air and so there is a minimum of summer precipitation.

In the fall and winter, the Pacific High weakens and migrates southward while the low-pressure center over the Aleutian Islands intensifies. The resulting circulation brings air from the south and southwest. This air, having had a long trajectory over warmer southern waters, arrives at the coast loaded with moisture. As the air moves inward and is forced up the windward slopes of the Olympic Mountains, it cools. Because the air is nearly saturated, the lifting causes precipitation quickly. At lower elevations, the precipitation is in the form of cold rain; but at about 2,000 feet, the cooling caused by lifting is intense enough to change the rain to snow. At about 4,500 feet, winter precipitation is nearly always in the form of snow. So not only does the amount of precipitation increase with elevation, but the portion in the form of snow increases. It is not surprising, therefore, that the higher elevations of the Olympics are spawning grounds for several large glaciers.

Because of the effect of the mountains in extracting moisture from the prevailing winds off the ocean, the role of migrating low-pressure storms and associated frontal systems is not as apparent as in the eastern part of the continent. Wintertime lows typically cross the Olympic Peninsula on a west-to-east track. The fronts associated with these lows are usually occluded and rather indistinct. Thus there is not the progression of weather events associated with typical frontal storms. Storm-associated precipitation increases and decreases slowly as the low passes by and may merge imperceptibly with the precipitation area associated with the next approaching storm.

Because the area is exposed to the sweep of winds from the Pacific, wind speeds tend to be high. The more intense winter storms frequently bring winds of near-hurricane force, and speeds of 50 mph occur every winter.

The interseasonal changes in circulation patterns are gradual, however, and not monsoonal in character, although the summer precipitation is typically only a fifth or less of the winter precipitation.

Solar data

Solar data for Mount Olympus are presented in Table 12-1. Winter days are about 8½ hours long; summer days, 16. Because

Table 12-1.   Solar data for Mount Olympus, Washington

Date	Sunrise	Solar noon	Sunset	Day length hr:min	Twilight min
Jan 1	0804	1218	1632	8:28	36
Jan 16	0759	1224	1650	8:51	35
Feb 1	0742	1228	1714	9:32	33
Feb 16	0720	1229	1738	10:18	32
Mar 1	0654	1227	1800	11:05	31
Mar 16	0625	1233	1822	11:57	31
Apr 1	0552	1218	1845	12:53	32
Apr 16	0522	1214	1906	13:44	33
May 1	0456	1212	1927	14:32	35
May 16	0434	1211	1948	15:13	38
June 1	0420	1213	2005	15:46	40
June 16	0415	1215	2016	16:00	41
July 1	0420	1219	2017	15:56	41
July 16	0433	1221	2008	15:35	39
Aug 1	0452	1221	1950	14:57	36
Aug 16	0512	1219	1926	14:13	34
Sept 1	0534	1214	1855	13:22	32
Sept 16	0554	1209	1825	12:31	31
Oct 1	0615	1204	1754	11:39	31
Oct 16	0636	1200	1724	10:48	31
Nov 1	0700	1158	1656	9:56	33
Nov 16	0723	1200	1636	9:13	34
Dec 1	0743	1204	1624	8:41	36
Dec 16	0758	1211	1623	8:25	36

Note: Pacific Standard Time in hours and minutes on 24-hour clock. Add one hour during Daylight Time. Mt. Olympus, Washington is Lat. 47°49′N Long. 123°40′W.

of the compact nature of the area, the data apply with small error to the entire region.

Bioclimatic index

The annual bioclimate progression depends on where you are with respect to the Olympic Mountains. At low elevations on the windward side of the range, the bioclimate ranges from cool/wet (very wet) in the winter and spring months to mild/humid from May through September (figure 12-2). As the precipitation increases in the fall, the bioclimate is progressively more humid, becoming mild/humid for a month or so before cooling off to typical winter values.

On the leeward side of the mountains (see figure 12-2), the bioclimate is markedly drier. Thus the bioclimate progresses from cool/humid from October through March to mild/dry from April through September.

At higher elevations, especially on the windward side of the mountains, the climogram would have the same general shape as that for Quillayute but displaced downward about 3 degrees for each thousand feet of increased elevation. Thus at Blue Glacier, high on the slopes of Mount Olympus at the timberline elevation of 6,000 feet, the climogram would be displaced about 20 degrees lower. Winter months would be placed in the cold/wet zone and summer months in the cool/humid zone. Above the timberline, the climogram would be displaced even farther downward, placing the winter months in zone II, very cold, or lower, depending on the wind speed. Even summer months may have a cold/humid bioclimate above the timberline.

The transition between the mild/humid weather of summer and the mild/wet weather of fall is gradual, with rainfall increasing steadily and temperatures decreasing moderately. At higher elevations, the transition will be from cool/humid in the summer to cool/wet in the fall. In the spring, the reverse transition is even more gradual; each month from January on becomes progressively warmer and less humid.

Summer

There is no sharp and definable beginning of the summer season. As the Aleutian Low weakens and the Pacific High moves northward, the prevailing winds gradually shift from southerly to westerly. This shift is accompanied by a gradual decline in the frequency and intensity of coastal storms. The moderating marine

Figure 12-2. Bioclimatic index

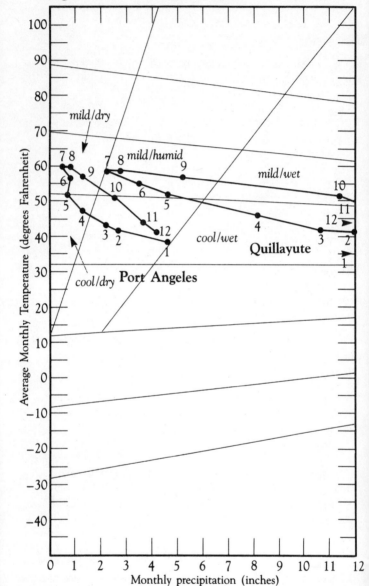

influence is strong and the monthly mean temperature increases only about 4 degrees per month from March to July. The daily range is also relatively small, about 16 to 18 degrees. In the lowlands along the west coast, the highest recorded temperatures are not much over 90°F for June through September (see figure 12-3). Farther inland, at Forks and Lake Quinalt, highest temperatures of record have been just over 100°F. On the north side of the peninsula, at Clallam Bay and Elwha Ranger Station, the highest temperatures have been in the mid-nineties. Most of the time, however, maximum daily temperatures during the summer months are near 70°F.

At higher elevations, afternoon temperatures typically range from 55°F to 65°F, occasionally reaching the lower seventies. Nighttime minimum temperatures range from the mid-fifties in low areas near the coast to the mid-thirties at higher elevations. Even at high elevations, summer minimum temperatures below freezing are comparatively infrequent.

The frost-free season is long at low elevations. Fifty percent of the time at Forks, the last date on which a minimum temperature of 32°F is reached is April 22; the earliest date for a freezing temperature is October 18. For a minimum temperature of 28°F and the same 50-percent probability of occurrence, the period is extended about a month in both spring and fall (see table 12-2). This means there is usually a long period during which the temperature does not go below freezing—an indication of the mild marine climate.

As might be expected, humidities are generally high—from 90 percent or more at sunrise to 50–65 percent in the afternoon. The lowest humidities (and the highest temperatures) occur when the wind is from the east. Warm, dry, easterly winds occasionally bring temperatures in the eighties and humidities as low as 20 or 30 percent.

Coastal fog is also common, especially in the late summer and early fall. These are really layers of stratus cloud that are formed over the ocean and move inland at night. The bottom of the stratus may be only a few hundred feet above sea level, with tops around 3,000 feet. At ground elevations within this range, the cloud appears as fog and saturates the foliage, causing so-called fog drip. The driest place to camp is thus in the open, for the steady capture of fog by the trees can deposit substantial quantities of water beneath the tree crowns. At elevations above the cloud top, typically 2,000–3,000 feet, the weather remains clear. Daytime sunshine generally dissipates the stratus by midday.

Figure 12-3. Temperature. Quillayute, Washington

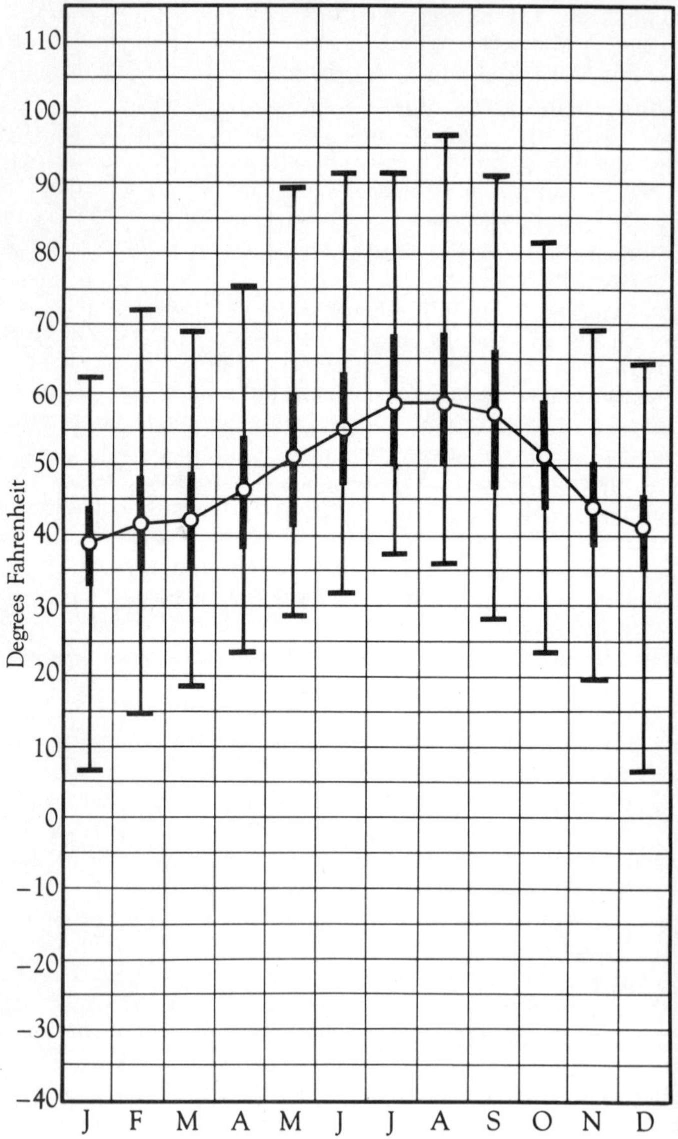

At low elevations, there are few completely sunny days in midsummer. About one day in four has predominantly clear weather during daylight hours (see figure 12-4). Partly cloudy conditions prevail for another fourth of the days; and cloudy or rainy weather accounts for the remaining half. Nearly all these days have some kind of precipitation, which, more often than not, will be drizzle. Above the stratus, however, it is not unusual for two or three weeks to pass with little more than an occasional shower. Thunderstorms are rare, although July and August average about one per month.

Precipitation amounts vary widely, depending on elevation and location. At low elevations along the Pacific coast, summer rainfall ranges from 2 to 4 inches per month; along the Strait of San Juan de Fuca, these amounts are cut in half (figure 12-5). Summer rain generally comes with winds from the south or southeast.

Rainfall intensities are generally low, too: drippy days are the rule, although the occasional shower or thundershower can produce brief downpours. Throughout the peninsula, July is the

Table 12-2. Probability of 32°F, 28°F, and 24°F occurring at Forks Ranger Station

Spring probability	75%	50%	25%	10%
32°F	Apr 9	Apr 22	May 6	May 18
28°F	Mar 6	Mar 19	Apr 1	Apr 13
24°F	—	Feb 15	Mar 4	May 17

Fall probability	10%	25%	50%	75%
32°F	Sep 25	Oct 6	Oct 18	Oct 30
28°F	Oct 21	Oct 31	Nov 13	Nov 26
24°F	Nov 10	Nov 23	Dec 12	—

Figure 12-4.   Sky and weather. Quillayute, Washington

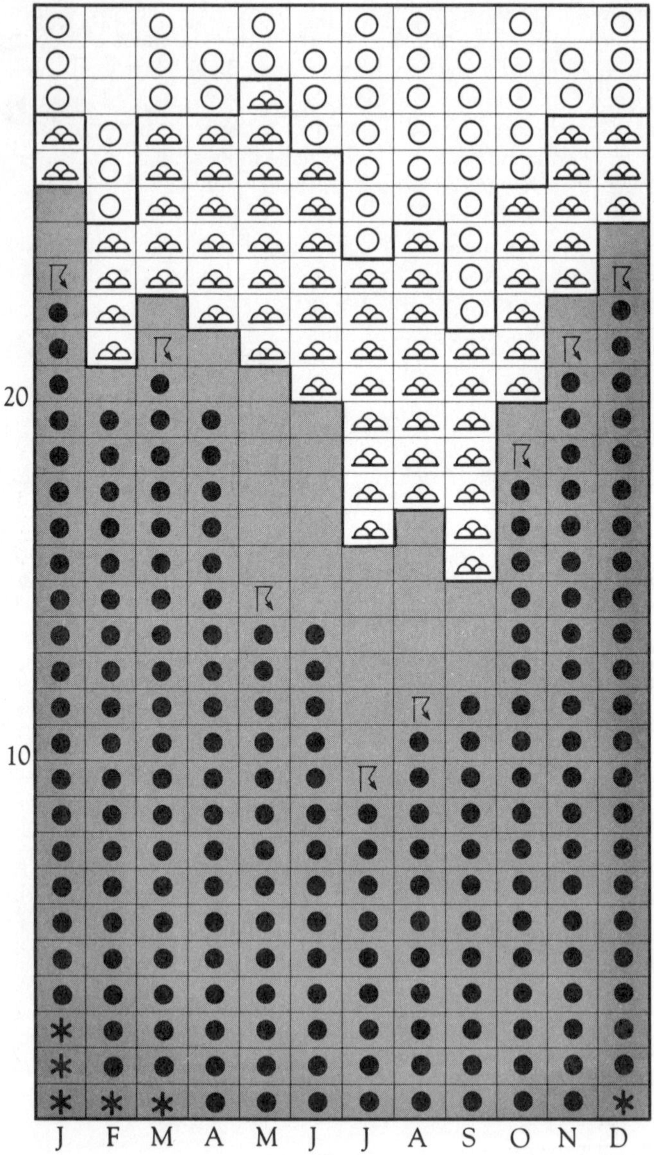

month with the least rainfall, although June and August have little more.

Wind speeds are moderate in the summer. At high elevations, daytime winds are typically from the west; at sea level, the direction is more from the southwest. Along the strait, the afternoon wind is generally a moderate westerly breeze.

Good rain gear is essential for the summer hiker. Although high elevations often have long stretches of predominantly fair weather, showers are frequent. And at lower elevations, the rain forest may drip almost continuously.

Figure 12-5. Monthly precipitation

Quillayute, Washington

Annual: 105 inches   Snowfall: 21 inches

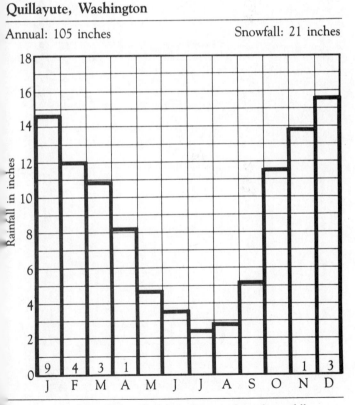

Figures at the base of the columns indicate inches of snowfall

Figure 12-5. *(continued)*

Port Angeles, Washington

Annual: 25 inches   Snowfall: 7 inches

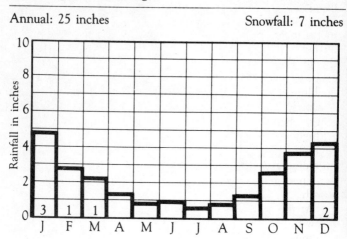

Elwha Ranger Station

Annual: 56 inches   Snowfall: 19 inches

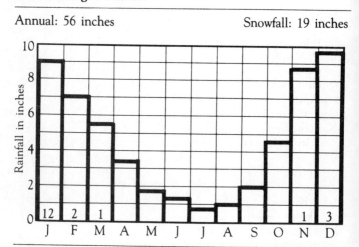

Figures at the base of the columns indicate inches of snowfall.

Autumn

The transition to autumn is gradual. The Aleutian Low intensifies and the Pacific High weakens and migrates southward. The result is a gradual intensification of a southwesterly flow of increasingly moist air. As the air moves up the slopes of the Olympics, it quickly reaches saturation and produces rain. Although the onset of the autumn rains is gradual, the largest increase in precipitation occurs from September to October, with a doubling of precipitation amounts. The number of rainy days and of cloudy days also increases markedly from September to October, when two-thirds of the days are predominantly overcast. At high elevations, rain and snow may alternate, although snow begins to predominate by late October. Heavy snowfalls have been observed as early as the first week in November.

There may still be an occasional thunderstorm, about one per month.

At the Forks Ranger Station—typical of valley locations—below-freezing temperatures are rare before October. Nighttime minimum temperatures of 24°F or lower are uncommon: the probability is greater than 50 percent that such a temperature will not be reached any time during the fall or winter (see table 12-2).

Winter

If the onset of winter is defined as the date of the first snowfall, then winter begins at the end of November in the low country and about two months earlier above the timberline. The last significant snowfall is generally in March along the coast and in late April in the high country. But this does not tell the whole story; at low elevations, the precipitation is nearly all rain and above the timberline, it is nearly all snow—a very wet and heavy snow. In December, January, and February, the monthly precipitation at Blue Glacier is about 20 inches of water equivalent, corresponding to about 7 feet of soggy snow each month. The total annual snowfall there, measured as water equivalent, is about 10 feet, equal to about 40 feet of snow. This, of course, is what keeps Blue Glacier going. But even in midwinter, some of the precipitation falls as rain.

In midwinter, the snowline lies between 1,500 and 3,000 feet above sea level. Above the timberline, snow covers the ground from November until June. At lower elevations, winter snows melt quickly and depths are rarely greater than 6 to 15 inches.

The rain shadow effect is very important in determining the distribution of winter precipitation. Because the moisture-bearing winds are from the southwest, there is a decreasing gradient of snowfall from southwest to northeast. The effect is less noticeable at higher elevations, but it is still present. For example, at Port Angeles, the January average is about 5 inches, compared with nearly 15 inches at Quillayute. At Elwha Ranger Station (elevation 345 feet), not far from Port Angeles, the January average is 8.7 inches; the January average at Forks, at the same elevation as Elwha, is 17.5 inches.

As might be expected from all this precipitation, winter days are cloudy. Three-quarters of the days are overcast, most of them with rain or snow. Rainfall intensities are generally moderate, although one storm in January 1935 produced 12 inches of rain in twenty-four hours at Lake Quinalt. This same storm produced 2 feet of rain in forty-eight hours and nearly 3 feet in four days. There is still the occasional thunderstorm in the winter months. Only about three days per month are sunny and clear; these are usually associated with outbreaks of cold, dry air from Canada or from the Cascades to the east.

Afternoon temperatures are generally in the forties and nighttime minimums are in the thirties at lower elevations. Occasional outbreaks of Canadian air bring lower temperatures, drier air, and sunny skies. Temperatures may then remain below freezing for a few days, with minimums around 10°F to 15°F. Such outbreaks may occur only once or twice a month and unfortunately do not last for more than a day or two.

Winter storms also bring high winds. Wind speeds measured in the open along the coast regularly exceed 50 mph in intense storms and have reached more than 90 mph. Wind data from an exposed site on a 2,000-foot ridge near the ocean indicate that speeds in excess of 100 mph occur in the higher elevations nearly every winter. Generally, however, wind speeds are less severe. The average speed at Quillayute is about 7–8 mph during winter months, and the highest speed recorded there during eleven years of observation was 46 mph in February 1977.

Spring

Spring does not really exist as a definable season in the Olympics. The transition from winter to summer is gradual as the circulation patterns shift ponderously from the regime dominated by the Aleutian Low to that dominated by the Pacific High.

Cloudiness, precipitation, and wind speed decrease slowly as temperatures rise. At low elevations, weather conditions may be appropriate for hiking in March and April, while higher elevations may still be in winter's grip. There, May precipitation is usually snow. High-country hiking may be difficult or impossible well into June.

Temperatures climb slowly in the spring. The probability is 50 percent that freezing temperatures will occur as late as April 22; and 10 percent that freezing temperatures will occur as late as May 18 (see table 12-2).

Summary

Hiking and backpacking are possible year-round at low elevations, although winter hikers will find rather wet and raw conditions. At higher elevations, winter is for mountaineers who are willing and prepared to cope with heavy, wet snowfall and day after day of clouds and precipitation. July is generally the best month for high-country hiking, with August not far behind. June is an iffy month in the high country because of the possible persistence of the snowpack. Any time of the year, good protection from rain is essential.

Additional information

A climatic summary containing additional information can be obtained from the National Climatic Data Center, Federal Bldg., Asheville, NC 28801:

> *Northwest Olympic Peninsula, Washington.* Climatography of the United States 20–45. U.S. Dept. of Commerce, Weather Bureau. 1963.

Blue Glacier, flowing northward from the upper slopes of Mount Olympus, has been studied extensively. For an introduction to this literature which contains considerable climatic data, see the following:

> LaChapelle, E. *The Mass Budget of Blue Glacier, Washington.* Journal of Glaciology 5(41):609–623. 1965.

Figure 13-1. The Sierra Nevada

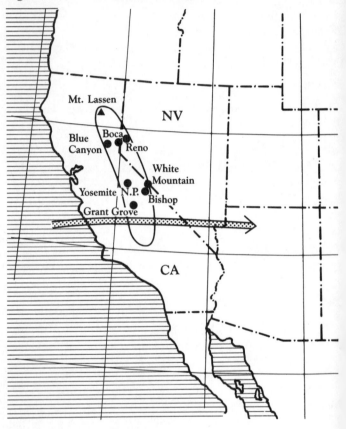

Arrow indicates winter storm track.

Chapter 13

The Sierra Nevada

The Sierra Nevada—Range of Light—stretch 400 mountainous miles from the southern terminus of the Cascades at Mount Lassen to Tehachapi in the south. From the west, the few roads that penetrate and cross the range start up from the near-desert of California's Central Valley and twist their way through the Sierra Nevada foothills. At first the trees are sparse, digger pines and live oaks scattered among the grassy hills. But soon the open woodlands give way to taller and statelier conifers: graceful ponderosa pine, fragrant sugar pine, and white fir. Higher still the giant pines give way to dense groves of red fir pushing their pointed tips nearly 200 feet in the air.

If one drives over the highest of the trans-Sierra Nevada passes, Tioga, nearly 10,000 feet high, the trees become sparser once again, small clumps separated by stretches of meadow or jumbles of granite. This subalpine zone is populated with pines: the bushy foxtail, the five-needle whitebark, and the occasional lodgepole, twisted and gnarled from exposure to timberline weather. Above the timberline, at about 11,000 feet, lies the true alpine zone; jumbled fields of granite boulders, sparkling tarns, and patches of high meadows.

Many forces have conspired over geologic time to form the spires, valleys, and lakes of the Sierra Nevada. Two of the most visible and spectacular of the forces are those that tilted the giant granitic block to the west, and the great valley glaciers that scoured out the valleys and put a high polish on much of the granite. The upheaval left the great ranges with a sharp, almost clifflike eastern edge. Nowhere is this scarp more impressive than at the southern end of Owen's Valley, where the crest of the ridge towers nearly 10,000 feet above the little town of Lone Pine, scarcely a dozen miles away. The ridge here culminates in the summit of Mount Whitney (14,495 feet), the highest point in the contiguous forty-eight states.

This great barrier to the eastward flow of winds from the Pacific has profound effects on the weather and climate, not only of the mountains themselves but of the rest of the continent to the east. As moisture-laden air is pushed up the western slopes, prodigious amounts of water are extracted. To the east, beyond the crest, the dried air descends to the intermountain valleys, producing near-desert conditions.

General climate

Despite the great height of the Sierra Nevada, the climate is surprisingly mild. The marine air currents that move inland on the prevailing southwesterly winds are warmed and wetted by their long trajectory over Pacific waters. Some of their wetness is diminished by their route over the low Coast Range, but enough is left to produce copious precipitation in the Sierra Nevada. As the moisture condenses, it releases heat, the same heat required to evaporate it from the ocean. As a result, winters, though snowy, are relatively mild. At Huntington Lake, the highest reporting station in the mountains at 7,020 feet, the average January temperature is 31°F. Aspen, Colorado, only a few hundred feet higher in elevation, has a January average 10 degrees lower. There is no climatological station at Tioga Pass; but the January mean there is probably about that of Aspen, although Tioga Pass is nearly 10,000 feet high.

The gradual western slope of the range greatly affects the distribution and amount of precipitation from winter storms. Below about 2,000 feet, winter precipitation is nearly always in the form of rain. Above this line, winter storms bring rain and snow in amounts that increase up to an elevation of 7,000 or 8,000 feet. The amounts decrease as the crest is approached because most of the water has been extracted at the lower elevations. On the east side, in the deep valleys along the California-Nevada line, the amount of precipitation declines sharply to

Figure 13-2. Monthly precipitation

Yosemite Valley, California

Annual: 36 inches   Snowfall: 73 inches

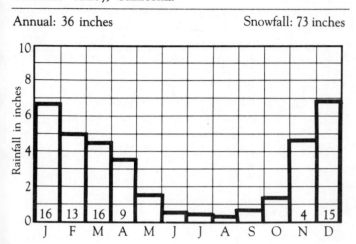

Bishop, California

Annual: 6 inches   Snowfall: 8 inches

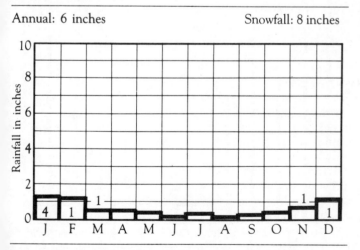

Figures at the base of the columns indicate inches of snowfall.

Figure 13-2 (continued)

Boca, California

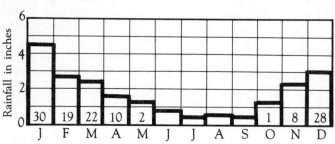

Annual: 22 inches   Snowfall: 120 inches

Blue Canyon, California

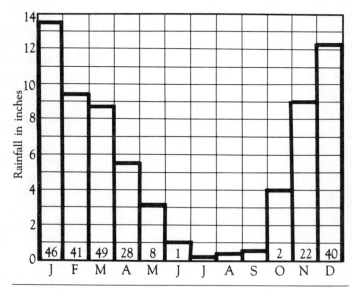

Annual: 68 inches   Snowfall: 237 inches

Figures at the base of the columns indicate inches of snowfall.

near-desert amounts (see figure 13-2). Snow displaces rain as the dominant winter precipitation at high elevations, although at Blue Canyon (elevation 5,280 feet), more than half the January precipitation is rain. The January mean is 13.7 inches, 6 inches of which produce an average 46 inches of snowfall. Higher up, winter precipitation is nearly all snow, although very often it is a rather heavy and soggy snow, not the light powder so characteristic of the Rockies.

In the summer, storm tracks are displaced far north. Low-pressure systems from the Pacific rarely invade the Sierra Nevada. High pressure dominates the inland mountains while a heat low occupies the Central Valley. Caught between these two fair-weather systems, the Sierra Nevada normally enjoys rain-free summers. Occasionally, however, moist air works its way up from the Gulf of Mexico and triggers a period of thundery and showery weather.

North-south climate variations are considerable because of the length of the range, extending over 5 degrees of latitude. Precipitation decreases substantially from north to south. Mean temperature is somewhat higher in the south, with the result that the southern end of the range is markedly milder and drier than the northern end. However, for the backpacker, these latitudinal differences are usually overshadowed by altitudinal changes: a thousand feet difference in elevation is the approximate climatic equivalent of the entire north-south extent of the Sierra Nevada.

The Sierra Nevada climate is a complex mosaic of altitudinal, latitudinal, and temporal climates. The analyses that follow focus on the high Sierra Nevada, 5,000 feet and above. Unfortunately, there are few climatic stations in this region. The highest climatological station reporting both temperature and precipitation is Huntington Lake (elevation 7,020 feet), although there are two high stations (elevations 10,150 feet and 12,470 feet) on White Mountain Peak east of Owens Valley. Thus the climate of the high peaks and passes and the above-timberline area must be inferred from lower elevation data.

Solar data

Tabulated data for Yosemite National Park are presented in Table 13-1. Because of its southern location, winter days are relatively long, nearly 10 hours; summer days are about 14½ hours. Because of the small size of the area, the tabulated data apply to the entire region with errors of less than ten minutes.

Table 13-1. Solar data for Yosemite National Park, California

Date	Sunrise	Solar noon	Sunset	Day length hr:min	Twilight min
Jan 1	0714	1202	1650	9:36	29
Jan 16	0712	1208	1704	9:51	29
Feb 1	0703	1212	1721	10:18	28
Feb 16	0647	1212	1738	10:50	27
Mar 1	0629	1211	1752	11:23	26
Mar 16	0607	1207	1807	12:00	26
Apr 1	0543	1202	1821	12:39	27
Apr 16	0521	1158	1835	13:14	27
May 1	0502	1155	1849	13:47	28
May 16	0447	1155	1902	14:15	30
June 1	0438	1156	1915	14:37	31
June 16	0436	1159	1922	14:46	32
July 1	0440	1202	1924	14:44	32
July 16	0450	1204	1919	14:30	31
Aug 1	0502	1205	1907	14:05	29
Aug 16	0515	1202	1850	13:35	28
Sept 1	0529	1158	1827	12:59	27
Sept 16	0541	1153	1804	12:23	26
Oct 1	0554	1148	1741	11:47	26
Oct 16	0608	1144	1719	11:11	27
Nov 1	0624	1142	1700	10:35	27
Nov 16	0640	1143	1646	10:06	28
Dec 1	0655	1148	1640	9:44	29
Dec 16	0708	1154	1641	9:33	30

Note: Pacific Standard Time in hours and minutes on 24-hour clock. Add one hour during Daylight Time. Yosemite National Park is Lat. 37°45′N Long. 119°35′W.

Bioclimatic index

Despite the height of the Sierra Nevada and their exposure to Pacific winds, the year-round bioclimate is mild. At Grant Grove in Sequoia National Park (elevation 6,600 feet), the winter bioclimate can be characterized as cool/wet (figure 13-3). Even on the crest, several thousand feet higher, the average bioclimate is only in the cold/wet zone. Wind speeds are generally moderate so that

Figure 13-3. Bioclimatic index

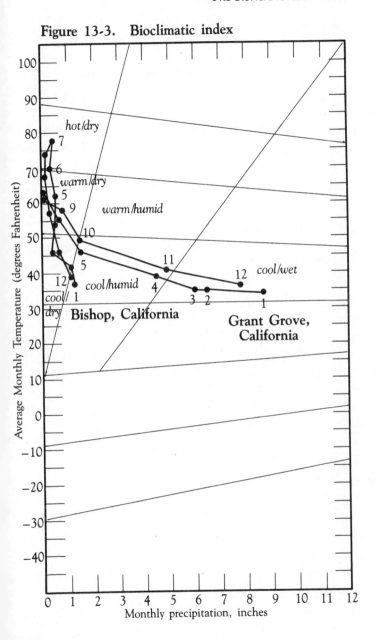

even during storms, the extremely bitter conditions that characterize above-timberline weather in the much lower Presidentials of the Northeast rarely occur.

In the summer, conditions are much drier. The bioclimate can be characterized as warm/dry at intermediate elevations and cool/dry at timberline.

The bioclimate east of the crest is dramatically different. Precipitation amounts are extremely low throughout the year: at Bishop, the average for January, the wettest month, is only 1.2 inches. And because of the dry climate, the summer-winter temperature fluctuation is much greater than on the west side of the crest. The Bishop bioclimate thus fluctuates from cool/dry in the winter to hot/dry in the summer.

These bioclimates are good news for the backpacker. The mild and dry summer weather is very nearly ideal for hiking, and the snowy and comparatively mild winters yield heavy snowpacks and generally good ski touring conditions.

Summer

If summer is defined as the dry season, then it lasts from May through October. If it is defined as the months when hiking is feasible in the high country, then summer usually lasts only from July through September. Although the weather earlier in the season may be fine for hiking, the deep winter snowpack may impede foot progress across the high passes.

Fifty percent of the days are overcast in the winter. The percentage drops steadily from April onward, until, in July, August, and September, fewer than 10 percent of the days are overcast and more than three-quarters of the days are predominantly clear. Typically, afternoon skies will be filled with puffy cumulus clouds that dissipate in late afternoon. Nights are usually cloudless, which promotes rapid radiational cooling and chilly nighttime temperatures.

Summer daytime temperatures are ideal for hiking. The average daily maximum at Grant Grove is 74–75°F in July and August, and the highest ever recorded there is 90°F (see figure 13-4). At higher elevations, daytime temperatures will be somewhat lower, but the abundant and intense sunshine tends to counteract the cool temperatures and keep the hiking climate comfortable.

Nighttime minimum temperatures develop in a mosaic that is much more complicated than daytime temperatures. Topography, elevation, and ground cover interact to produce microclimatic variations that may produce differences in minimum temperatures

Figure 13-4. Temperature. Grant Grove, California

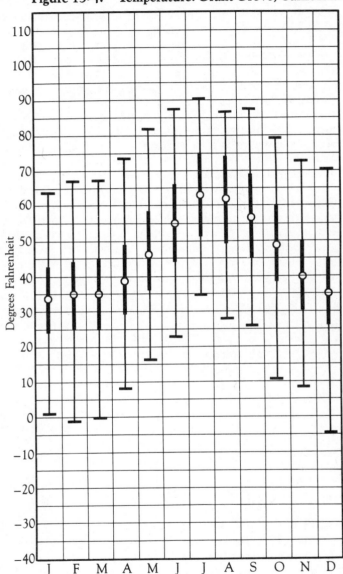

of 10 or 20 degrees in just a few hundred yards. Although freezing temperatures are rare at intermediate elevations, especially in the timber (the lowest recorded minimum at Grant Grove in July was 35°F), high meadows lying in broad basins may collect cold air cascading from nearby slopes and be true frost pockets. Here skims of ice on the water bucket are not uncommon in the early morning. Warm nights are rare; and typical nighttime temperatures are in the forties.

For the backpacker, one of the delights of the summer Sierra Nevada is the almost complete absence of rain. At Grant Grove, the number of days with more than 0.1 inch of precipitation is one or less for the entire summer—June through September (see figure 13-5). Days with 0.5 inch or more are almost unknown. Nevertheless, rainy days do occur. I remember with some embarrassment one July in the mountains above Tuolumne Meadows. I was introducing some friends to the delights of the high Sierra Nevada and had promised "it never rains in the summer." Although each day dawned bright and clear, clouds built up by late morning and persistent showers sputtered rain all afternoon and well into the evening for five consecutive days. Fortunately, we had plastic tarps and ponchos, so we kept dry. But my credibility as a weatherman suffered an irreparable blow.

Figure 13-5.　Precipitation frequency. Grant Grove, California

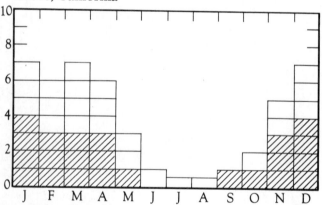

Number of days per month with 0.1 inches of precipitation or more. Hatched areas indicate days with 0.5 inches precipitation or more.

Figure 13-6. Sky and weather. Blue Canyon, California

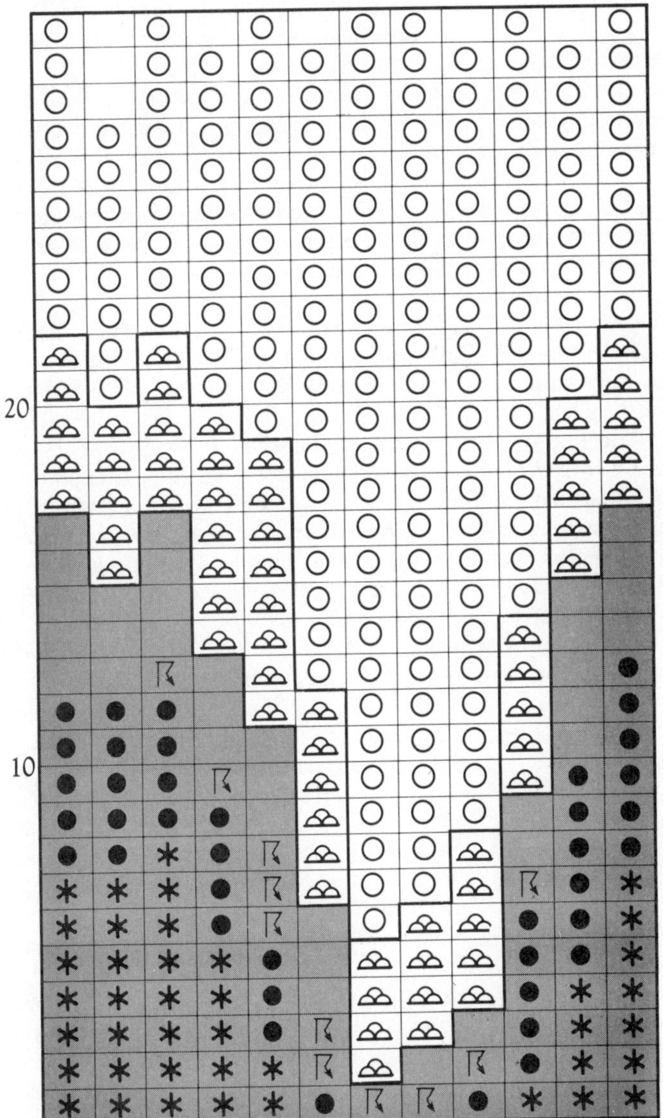

Sierra Nevada hikers in early September 1978 did not fare so well. Tropical storm Norman brought rain and low temperatures to the southern mountains and snow above the timberline. Many hikers, caught by surprise, suffered hypothermia and four died. Although this was a most unusual storm, the lesson is clear: hikers venturing above the timberline must be aware of the possibility of severe weather any time. This does not mean everyone must carry full winter regalia in midsummer. It does mean hikers must be ready to modify their routes and timetables if life-threatening weather should develop. It is better to hole up in the timber, where protection and firewood are available, than to perish trying to cross a snowed-in pass.

But storms of this sort are rare in the summer. Even thunderstorms do not occur frequently. Two or three thunderstorm days per month is the norm for Blue Canyon (see figure 13-6). In the high country, this figure may double; towering cumulus clouds build up over the peaks. On the east side, in Owens Valley, July thunderstorms are more frequent; but there is usually not enough moisture in the air over the mountains for these clouds to produce much (see figure 13-7). However, when high-level currents bring moisture from the Gulf of California or the Gulf of Mexico, showers and thundershowers break out. Such an occurrence is usually heralded by high-level clouds moving in from a southerly or southeasterly direction.

The more typical westerly winds may also bring in high-level moisture. The most spectacular cloud form of the Sierra Nevadas, the lenticular banner clouds that cap the highest peaks, develop when the jet stream sits athwart the range. As the high-level air is forced upward, moisture condenses and forms a cloud. As the air descends on the lee side of the range, the cloud evaporates as the descending air is compressed and heated adiabatically. The result is a stationary cloud, sometimes appearing as an upside-down pile of shallow bowls. This up-and-down motion sometimes causes secondary bounces in the high air, forming additional streets of lenticular clouds perpendicular to the direction of the streaming air. This is the famous Sierra Wave. Glider pilots and hang-gliders seek the upward-moving air, although it has its dangers. Gliders have been known to be caught in such swift updrafts that the pilot could not bring them down before blacking out from lack of oxygen.

Despite the frequent presence of these fast-moving, high-level currents, surface winds in the Sierra Nevada tend to be moderate, even above the timberline. At Blue Canyon, well

Figure 13-7. Sky and weather. Bishop, California

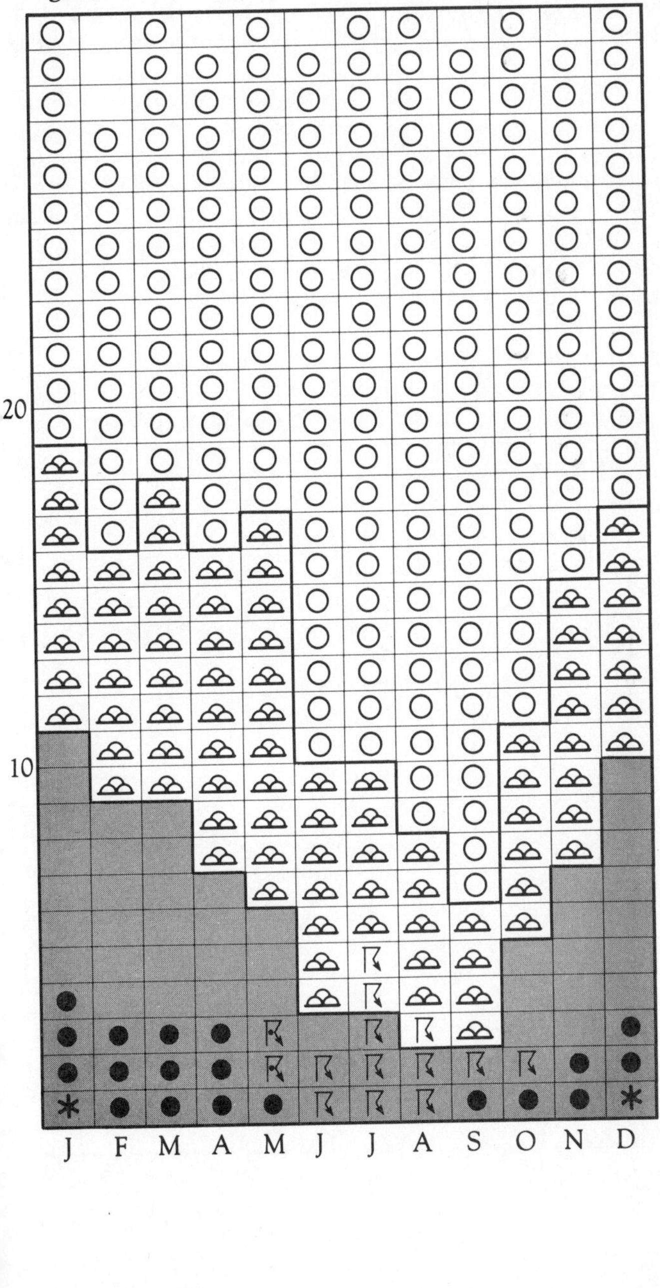

below the timberline, the average summertime winds are only 7 or 8 mph. More significantly, the maximum recorded winds have been around 40 mph. At the crest, of course, wind speeds will be somewhat higher.

Ideal backpacking weather is thus almost unbroken during the summer months. Nevertheless, nights are cool, even cold; and a warm sleeping bag is essential. Hardy hikers may venture into the mountains with minimal shelter: tarps or tube tents. But adequate protection from possible cold rain is essential: wool hat and gloves, trousers, warm sweater or parka, and adequate rain covering. During early and late summer, the possibility of a snowstorm must be considered, especially above the timberline.

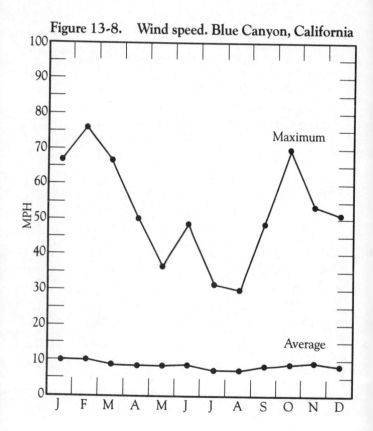

Figure 13-8. Wind speed. Blue Canyon, California

Autumn

Fall is a delightful time in the Sierra Nevada, but somewhat risky in the high country. Long stretches of bright and clear October sunshine may end abruptly with a 2-foot snowstorm, or good hiking weather may persist into November. At Blue Canyon, at a relatively low elevation, as much as 2 feet of snow have fallen in October and 6 feet in November. More often than not, however, there is little or no snow in October, even in the high country. November nearly always sees the onset of heavy snows.

Storm frequency increases as storm tracks move south for the winter. At Grant Grove, there is an average of two days in October with at least 0.1 inch of precipitation. Although this may not seem very much, it is up from one or two such days for the entire summer season. Cloudiness also increases as moisture moves in from the Pacific. Three-quarters of September days are clear, but the percentage drops to near fifty in October and thirty-three in November.

Temperatures drop rapidly as well. Daytime maxima fall to the fifties in the high country; nighttime minima will usually be below freezing. Wind speed increases to a secondary maximum in October.

In October, there is 95 percent probability of less than 0.1 inch of precipitation in any two-day stretch. This is a decrease from the 99-percent probability in September. In November, the probability drops to about 90 percent.

In every way, October weather is often ideal for late season backpacking; but equipment should be appropriate for the increasing likelihood of being caught by an early snowstorm or at least a cold rainstorm.

Winter

"The first general winter storm that yields snow that is to form a lasting portion of the season's supply, seldom breaks on the mountains before the end of November." So wrote John Muir in his journal, *The Mountains of California*, nearly a hundred years ago. "The first heavy fall is usually from two to four feet in depth. Then, with intervals of splendid sunshine, storm succeeds storm, heaping snow on snow, until thirty to fifty feet has fallen. But on account of its settling and compacting, and the almost constant waste from melting and evaporation, the average depth found at any time seldom exceeds ten feet in the forest region, or fifteen feet along the slopes of the summit peaks." Ten feet is a lot of snow, as several of us learned when visiting the Sierra Club's Claire Tappaan Lodge near Donner Summit. Our share of the chores was to dig out a 20-foot ladder the lodge manager needed. It had been left lying on the ground the previous fall. Below the halfway level, we had to relay the snow up to the surface, the bottom man dumping the snow on an intermediate platform, the next shoveling it to the surface, and the third removing it from the top to make way for more.

Muir and the shovelers could agree on another thing: Sierra Nevada snow is heavy. New-fallen density is typically about 12 to 15 percent, that is, it takes only 7 or 8 inches to melt to 1 inch of water. In the Rockies, fresh new powder usually has a density of 7 or 8 percent. There the melt ratio is more like 12 or 14 inches of snow to one of water. But Pacific storms are relatively warm and heavily laden with moisture. Air temperatures during winter storms are rarely much below freezing, even at higher elevations.

The elevation of persistent winter snow is about 4,000 feet on the western slopes of the Sierra Nevada. Snowfall increases with elevation, reaching a maximum at about 7,000 to 8,000 feet. Higher elevations receive less snowfall because most of the moisture in the storms advancing up the slope has been removed. Over the crest, snowfall amounts decrease even more. Blue Canyon, near the level of maximum snowfall, has an average of 20 feet; at Boca, at the same elevation but just east of the summit, the annual total is just half that. Farther east at Reno, the year's total is down to 2 feet; and at Bishop, it is only 9 inches.

Snowfall also decreases from north to south. In Yosemite Valley, 120 miles south of Blue Canyon, the winter snowfall is 6 feet. However, the valley is about 1,300 feet lower than Blue Canyon and so the lower amount at Yosemite is partly the result of its lower elevation. I estimate the total snowfall in Yosemite Park at an elevation comparable to Blue Canyon would be about 13 feet; and in Sequoia National Park, another hundred miles farther south, the amount would be about 10 feet.

There is considerable year-to-year variation in snowfall totals. In thirty-five years of record at Blue Canyon, winter snowfall amounts have ranged from 87 inches (in 1946–47) to 526 inches (in 1951–52). In January 1952, the average daily snowfall was 6 inches. By contrast, January four years earlier had only a trace of snow. Winter came a little late that year, but it came nevertheless. Snowfalls of 43 inches, 90 inches, and 119 inches in the following three months helped bring the winter total up to 303 inches. That year it even snowed 2 feet in May.

With all that snow, it might be expected that the Sierra Nevada would have a severe avalanche problem. Generally that is not so. Because of the dense, wet snow and the relatively mild winter temperatures, snowpacks consolidate well and tend to be avalanche-free, except for brief periods after winter storms. A comprehensive compilation of avalanche accidents in the United States from 1967 to 1971 showed only three out of seventy-six occurred in California. Ski areas that have an avalanche problem, such as Mammoth Mountain, have an active avalanche-control program. Of course, backcountry skiers should be alert to the terrain and snow conditions conducive to avalanche formation. Sluffing is common during winter storms and soft-slab avalanches may occur, particularly if the storm is followed by cold weather.

Storm probability increases during the winter months. Although November has only about a 5-percent probability that more than 0.1 inch of precipitation will occur in the next forty-eight hours, the probability increases to about 10 percent in December and 15 percent in January and February. Put another way, an average of six or seven days each winter month will have 0.1 inch or more of precipitation. About half these days will have major storms with more than 0.5 inch of precipitation, equivalent to 5 or 6 inches of snowfall. An occasional curious accompaniment to a winter storm is thunder and lightning.

Winter temperatures are comparatively mild. At Grant Grove (elevation 6,600 feet), the January average is only 32°F and the absolute minimum temperatures are near 0°F for December through March. Occasional winter warm spells bring the daytime maxima into the sixties. Temperatures are, of course, lower at higher elevations, decreasing about 3 degrees for every additional thousand feet.

Wind speeds are moderate. The all-time maximum at Blue Canyon was 76 mph in January 1974, but the winter average is only 8–10 mph. Only an occasional winter storm is accompanied by high winds.

Thus the winter climate is normally propitious for snow sports. Heavy snow almost always blankets the Sierra Nevada by mid-December. Storms typically last from two to five days, with intervals of one to two weeks of dry weather. Ski tours started at the end of a winter storm are therefore likely to enjoy ideal conditions for a substantial period of time. Touring during a storm, however, is likely to involve breaking trail through deep fresh snow, an exhausting procedure. Equipment for moderate winter weather is required.

Spring

May is the transition month in the Sierra Nevada. A third of the time, May will have one or more major winter storms; on another third, there will be no fresh snow at all except at the highest elevations. Snowfall in June is rare; the summer dry season is in full swing. A third of all Junes will have less than 0.1 inch of precipitation; the wettest June on record at Blue Canyon saw only 3 inches of rain.

Springtime sees a maximum in the number of thunderstorms as daytime heating coupled with a still-adequate supply of moisture conspire to produce towering cumulus clouds and thunderheads. About three days in May and two in June have thunderstorms at Blue Canyon. At higher elevations, the number will be slightly larger.

Despite the lingering winter storms and the occasional thunderstorm, the number of clear days increases dramatically from twelve in May to eighteen in June. With long periods of sunshine, daytime temperatures climb, with the normal maximum up to 66°F in June at Blue Canyon. Winds diminish also. With the ending of winter storms, the maximum recorded wind speeds drop to around 40 mph.

The major impediment to travel is the lingering snowpack. Foot travel through the high country is usually impossible in May and iffy in June. Ski touring in the high country is usually possible in April and early May, but extremely variable snow conditions make waxing tricky. I can remember one spring tour to Ostrander Hut in Yosemite Park when no wax we had with us could prevent great globs of sticky snow from gluing itself to our skis. Unfortunately, the snow was still too deep and mushy to permit walking. So our skis soon became as heavy as our packs. Despite these occasional hazards, spring can be a delightful time in the high country; bright, warm sunshine prevails most of the time.

Summary

The dominant feature of the Sierra Nevada recreation climate is the sharp contrast between wet winters and dry summers. From November through April, precipitation averages six or seven times more than the other six months. The four summer months are very dry indeed. At Blue Canyon, June, July, August, and September have a total of 1.9 inches of rain as compared with 68 inches for the entire year. The winters are characterized by cloudy weather and frequent winter storms.

The long dry summers provide superb hiking weather. Rain is so infrequent that many hikers carry only minimal rain gear and light-weight tents. But nights can be cold above timberline and in places subject to cold-air drainage; a warm sleeping bag is appropriate. Although wind speeds are not excessive, even above timberline, the protection of a parka is often needed.

Winters are snowy at elevations above about 5,000 feet, although rain may mix in any time. Skiing is usually good; only an occasional year has a sparse snow cover. Winter temperatures are relatively mild but the backcountry ski-tourer will need protection against heavy snow and moderate cold.

The transition months of October and April herald the rapid change between the dominant seasons of winter and summer. The change is often rather abrupt, with the onset of winter heralded by a late October or early November snowstorm. Springtime transition may be nearly as abrupt, but a persistent snowpack may make it look like winter well into June or even July.

Additional information

John Muir was a perceptive observer and colorful journalist of the weather. Some of the best anecdotal information of the weather and climate of the Sierra Nevada are contained in his writings. See especially:

Muir, John. *The Mountains of California.* New York: The Century Company. 1898.

A recent scientific conference was devoted entirely to the weather and climate of the Sierra Nevada. A copy of the proceedings can be purchased from the American Meteorological Society, 45 Beacon Street, Boston, MA 02108.

Index